HYDRAULIC AND COMPRESSIBLE FLOW
TURBOMACHINES

Hydraulic and compressible flow turbomachines

A. T. Sayers, BSc (Eng), MSc, PhD, C Eng, MIMechE
Department of Mechanical Engineering
University of Cape Town

McGraw-Hill Book Company

London · New York · St Louis · San Francisco · Auckland · Bogotá
Guatemala · Hamburg · Lisbon · Madrid · Mexico · Montreal
New Delhi · Panama · Paris · San Juan · São Paulo · Singapore · Sydney
Tokyo · Toronto

Published by

McGRAW-HILL Book Company (UK) Limited

Shoppenhangers Road, Maidenhead,
Berkshire, England, SL6 2QL.
Telephone Maidenhead (0628) 23432
Cables MCGRAWHILL MAIDENHEAD Telex 848484
Fax 0628 35895

British Library Cataloguing in Publication Data
Sayers, A. T. (Anthony Terence), *1946–*
  Hydraulic and compressible flow turbomachines.
  1. Turbomachinery
  I. Title
  621.8

  ISBN 0-07-707219-7

Library of Congress Cataloging-in-Publication Data
Sayers, A. T., 1946–
  Hydraulic and compressible flow turbomachines / A. T. Sayers.
    p.   cm.
  Includes bibliographical references (p   )
  ISBN 0-07-707219-7
  1. Turbomachines.   I. Title
TJ267.S26   1990   89-13688
621.406—dc20   CIP

12345   93210

Typeset by Thomson Press (India) Limited, New Delhi
and printed and bound in Great Britain by Page Bros (Norwich) Ltd

To
HANNAH
WILLIAM
GILES

CONTENTS

PREFACE

This book has arisen from a collection of lecture notes compiled for an undergraduate course in turbomachines. The subject of turbomachines covers both hydraulic and compressible flow machines, and, while many specialized books concerning specific types of machines are available, the author was not able to find a book covering both hydraulic and compressible flow machines at a suitable level.

The book is aimed at the undergraduate and diploma student, and introduces terms and concepts used in turbomachinery. It is not an exhaustive text on the subject, nor is it intended as a design text, although many design parameters are given. It should rather be used as an introduction to the more specialized texts. The book assumes that the student has followed a course of basic fluid mechanics and thermodynamics, but mathematical manipulations are minimal. Each chapter attempts to be self-contained with regard to a particular type of machine, and it is suggested that the lecturer supplements the text with worked examples during the lecture and tutorial sessions, when the maximum understanding of the subject is derived. To aid in this, some worked exercises are supplied at the end of each chapter.

In a book of this nature, great patience is required in the typing and correcting of the manuscript. In this regard I would like to convey my great appreciation to Mrs Iris von Bentheim and Mrs Lyn Scott, who jointly performed this task. For photographic reproduction, I would like to thank Mr Vernon Appleton, while for much of the draughting of the diagrams, I thank my wife, Susan, without whose constant encouragement this textbook would not have been written. Finally, my thanks must go to the many students who have over the years helped to correct the notes and tutorial problems during their undergraduate studies.

A.T.S.

SYMBOLS

A	area	m^2
a	velocity of sound	m/s
b	depth or height of blade	m
C	absolute velocity	m/s
C_D	drag coefficient	
C_{DA}	annulus drag coefficient	
C_{DC}	tip clearance drag coefficient	
C_{DS}	secondary loss drag coefficient	
C_L	lift coefficient	
C_p	specific heat at constant pressure	J/kg K
C_s	spouting velocity	m/s
C_v	nozzle velocity coefficient	
c	blade chord	
D	diameter	m
	drag force	N
d_H	hydraulic diameter	m
E	power per unit weight of flow	W/(N/s)
e	radius of inscribed circle	m
F	force	N
g	gravitational acceleration	m/s^2
H	head	m
h	head loss	m
	specific enthalpy	J/kg
	blade height	m
I	impeller constant	J/kg
i	incidence angle	deg

L	lift force	N
	length	m
l	span	m
M	Mach number	
	mass	kg
m	mass flow rate	kg/s
N	number of stages	
	speed	rpm
$NPSH$	net positive suction head	m
N_s	dimensionless specific speed	rev or rad
N_{sp}	dimensionless power specific speed	rev or rad
P	power	W
$\bar{P}$	power coefficient	
p	pressure	Pa
Q	heat transfer to system	W
	volume flow rate	m^3/s
q	specific heat transfer	J/kg
	leakage flow rate	m^3/s
R	radius	m
	eye tip radius	m
	gas constant	J/kg K
Re	Reynolds number	
r	radius	m
	eye hub radius	m
SP	specific speed	rad/s
s	blade pitch	m
	specific entropy	J/kg K
T	temperature	K
	time	s
	torque	N m
U	tangential velocity	m/s
u	internal energy per unit mass	J/kg
V	velocity	m/s
v	volume	m^3
	velocity	m/s
W	work done per second	W
	relative velocity	m/s
Y	pressure loss coefficient	
Z	number of blades	
	elevation or potential head	m

Greek

α	angle of absolute velocity vector	deg
β'	angle of relative velocity vector	deg

β	blade angle	deg
Γ	circulation	m^2/s
γ	ratio of specific heats	
δ	deviation	deg
ε	deflection	deg
ζ	enthalpy loss coefficient	
	stagger angle	deg
η	efficiency	
θ	camber angle	deg
λ	work done factor	
μ	absolute viscosity	Pa s
ρ	density	kg/m^3
σ	cavitation parameter	
	solidity ratio	
σ_c	critical cavitation parameter	
σ_s	slip factor	
ϕ	flow coefficient	
ψ	power input factor	
	stage loading coefficient	
	head coefficient	
ω	angular velocity	rad/s

Subscripts

a	axial
atm	atmospheric
av	average
b	blade
c	casing
	compressor
D	design condition
d	draft tube
f	friction
fp	pipe friction
g	gate
H	hydraulic
h	hub
i	impeller
	inlet
l	leakage
m	at mean diameter
	mechanical
N	nozzle
nom	nominal

o	outlet
	overall
	total (stagnation) conditions
p	projected area
R	rotor
	at tip
r	radial
r	runner
	at hub
rel	based on relative velocity
s	isentropic
	shaft
	static
suc	suction
t	tip
	turbine
t–s	total-to-static
t–t	total-to-total
u	unit quantities
v	volumetric
vap	vapour
x	tangential

ONE

INTRODUCTION

1.1 DEFINITION

A turbomachine can be described as any device that extracts energy from or imparts energy to a continuously moving stream of fluid, the energy transfer being carried out by the dynamic action of one or more rotating blade rows. The dynamic action of the rotating blade rows sets up forces between the blade row and fluid, while the components of these forces in the direction of blade motion give rise to the energy transfer between the blades and fluid. By specifying that the fluid is moving continuously, a distinction is drawn between the turbomachine and the positive displacement machine. In the latter, the fluid enters a closed chamber, which is isolated from the inlet and outlet sections of the machine for a discrete period of time, although work may be done on or by the fluid during that time. The fluid itself can be a gas or a liquid, and the only limitations that we shall apply are that gases (or steam) are considered perfect and that liquids are Newtonian.

The general definition of the turbomachine as used above covers a wide range of machines, such as ship propellers, windmills, waterwheels, hydraulic turbines and gas turbines, and is therefore rather loose for the purposes of this text. We will limit ourselves to a consideration of only those types of turbomachines in which the rotating member is enclosed in a casing, or shrouded in such a way that the streamlines cannot diverge to flow around the edges of the impeller, as would happen in the case of an unshrouded windmill or aerogenerator.

The types of machines falling into our defined category and which will be considered in detail in succeeding chapters are listed in Table 1.1 and fall into

Table 1.1 Types of turbomachines

Turbomachines in which	
Work is done *by* fluid	Work is done *on* fluid
Axial flow hydraulic turbine	Centrifugal pump
Radial flow hydraulic turbine	Axial flow pump
Mixed flow hydraulic turbine	Centrifugal compressor
Axial flow gas turbine	Axial flow compressor
Pelton wheel hydraulic turbine	Radial flow fan

one of two classes depending on whether work is done by the fluid on the rotating member or whether work is done by the rotating member on the fluid. Types of turbomachines can also be defined as to the manner of fluid movement through the rotating member. If the flow is essentially axial with no radial movement of the streamlines, then the machine is classed as an axial flow machine; whereas if the flow is essentially radial, it is classed as a radial flow or centrifugal machine. Other special types of turbomachines exist, e.g. the Minto wheel or Baki turbine, but they will not be considered in this text.

Considering the two classes of machines listed in Table 1.1, some broad generalizations may be made. The first is that the left-hand column consists of machines in which the fluid pressure or head (in the case of a hydraulic machine) or the enthalpy (in the case of a compressible flow machine) decreases from inlet to outlet, whereas in the right-hand column are listed those machines which increase the head or enthalpy of the fluid flowing through them. This decrease or increase in head, when multiplied by the weight flow per unit time of fluid through the machine, represents the energy absorbed by or extracted from the rotating blades, which are fixed onto a shaft. The energy transfer is effected in both cases by changing the angular momentum of the fluid. It might therefore be reasonable to assume that different types of turbomachine would exhibit differing shapes of blades and rotating members, and this indeed is the case, as is shown in Fig. 1.1. In addition, because turbomachines have developed historically at different times, names have been given to certain parts of the machines as well as to different types of machines, and these are now defined.

Turbine. A machine that produces power by expanding a continuously flowing fluid to a lower pressure or head; the power output is usually expressed in kW.

Pump. A machine that increases the pressure or head of a flowing liquid, and which is usually expressed in kPa or m.

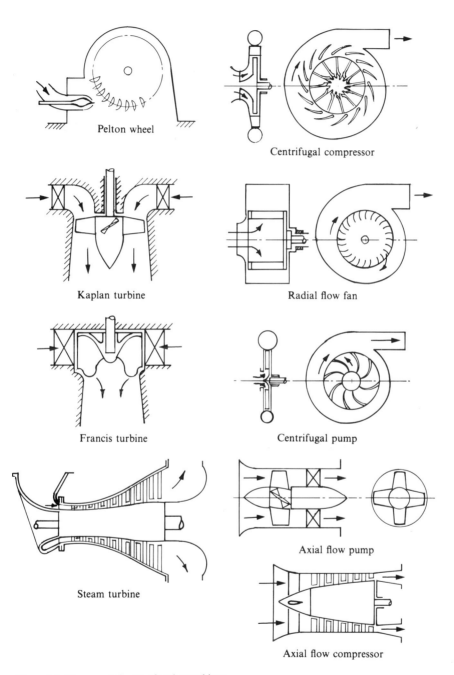

Pelton wheel

Centrifugal compressor

Kaplan turbine

Radial flow fan

Francis turbine

Centrifugal pump

Steam turbine

Axial flow pump

Axial flow compressor

Figure 1.1 Types and shapes of turbomachines

Fan. A term used for machines imparting only a small pressure rise to a continuously flowing gas, usually with a density ratio across the machine of less than 1.05 such that the gas may be considered to be incompressible; pressure increase is usually expressed in mm of water.

Compressor. A machine imparting a large pressure rise to a continuously flowing gas with a density ratio in excess of 1.05.

Impeller. The rotating member in a centrifugal pump or centrifugal compressor.

Runner. The rotating member of a radial flow hydraulic turbine or pump.

Rotor. The rotating member of an axial flow gas or steam turbine; sometimes called a disc.

Diffuser. A passage that increases in cross-sectional area in the direction of fluid flow and converts kinetic energy into static pressure head; it is usually situated at the outlet of a compressor.

Draught tube. A diffuser situated at the outlet of a hydraulic turbine.

Volute. A spiral passage for the collection of the diffused fluid of a compressor or pump; in the hydraulic turbine the volute serves to increase the velocity of the fluid before entry to the runner.

1.2 UNITS

The units used will be those of the SI system, the basic units being the kilogram, metre and second.

1.3 DIMENSIONAL ANALYSIS

The large number of variables involved in describing the performance characteristics of a turbomachine virtually demands the use of dimensionless analysis to reduce the variables to a number of manageable dimensional groups. Dimensional analysis also has two other important uses: firstly, the prediction of a prototype performance from tests conducted on a scale model; and secondly, the determination of the most suitable type of machine on the basis of maximum efficiency for a specified range of head, speed and flow rate. Only a brief description of the method used for forming dimensionless groups and their application to model testing for turbomachines will be given here, the generalities of the subject usually being covered in a course of fluid mechanics.

1.3.1 Hydraulic Machines

Figure 1.2 shows a control volume through which an incompressible fluid of density ρ flows at a volume flow rate of Q, which is determined by a valve opening. The head difference across the control volume is H, and if the control

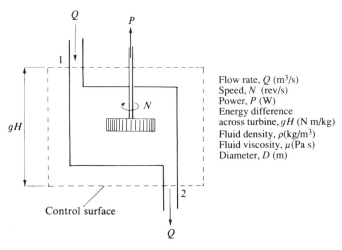

Flow rate, Q (m³/s)
Speed, N (rev/s)
Power, P (W)
Energy difference
across turbine, gH (N m/kg)
Fluid density, ρ(kg/m³)
Fluid viscosity, μ(Pa s)
Diameter, D (m)

Figure 1.2 Hydraulic turbine control volume

volume represents a turbine of diameter D, which develops a shaft power P at a speed of rotation N, then we could say that the power output is a function of all the other variables, or

$$P = f(\rho, N, \mu, D, Q, (gH)) \qquad (1.1)$$

In Eq. (1.1), f means 'a function of' and g, the acceleration due to gravity, has been combined with H to form the energy per unit mass instead of energy per unit weight. We now assume that Eq. (1.1) may be written as the product of all the variables raised to a power and a constant, such that

$$P = \text{const}[\rho^a N^b \mu^c D^d Q^e (gH)^f] \qquad (1.2)$$

   If each variable is expressed in terms of its fundamental dimensions, mass M, length L and time T, then, for dimensional homogeneity, each side of Eq. (1.2) must have the same powers of the fundamental dimensions, so the indices of M, L and T can be equated to form a series of simultaneous equations. Thus

$$(\text{ML}^2/\text{T}^3) = \text{const}(\text{M}/\text{L}^3)^a(1/\text{T})^b(\text{M}/\text{LT})^c(\text{L})^d(\text{L}^3/\text{T})^e(\text{L}^2/\text{T}^2)^f \qquad (1.3)$$

and equating the indices we get

M $1 = a + c$
L $2 = -3a - c + d + 3e + 2f$
T $-3 = -b - c - e - 2f$

There are six variables and only three equations. It is therefore possible to solve for three of the indices in terms of the remaining three. Solving for a, b

and d in terms of c, e and f we get

$$a = 1 - c$$
$$b = 3 - c - e - 2f$$
$$d = 5 - 2c - 3e - 2f$$

Substituting for a, b and d in Eq. (1.2),

$$P = \text{const}[\rho^{1-c} N^{3-c-e-2f} \mu^c D^{5-2c-3e-2f} Q^e (gH)^f]$$

and collecting like indices into separate brackets,

$$P = \text{const}[(\rho N^3 D^5)(\mu/\rho N D^2)^c (Q/N D^3)^e (gH/N^2 D^2)^f] \qquad (1.4)$$

The second term in the brackets will be recognized as the inverse of the Reynolds number and, since the value of c is unknown, this term can be inverted and Eq. (1.4) may be written as

$$P/\rho N^3 D^5 = \text{const}[(\rho N D^2/\mu)^c (Q/N D^3)^e (gH/N^2 D^2)^f] \qquad (1.5)$$

Each group of variables in Eq. (1.5) is truly dimensionless and all are used in hydraulic turbomachinery practice. Because of their frequent use, the groups are known by the following names:

$$P/\rho N^3 D^5 = \bar{P} \qquad \text{the power coefficient}$$
$$Q/N D^3 = \phi \qquad \text{the flow coefficient}$$
$$gH/N^2 D^2 = \psi \qquad \text{the head coefficient}$$

The term $\rho N D^2/\mu$ is equivalent to the Reynolds number $Re = \rho VD/\mu$, since the peripheral velocity V is proportional to ND. Hence Eq. (1.1) may be rewritten as

$$\bar{P} = f(Re, \phi, \psi) \qquad (1.6)$$

which states that the power coefficient of a hydraulic machine is a function of Reynolds number, flow coefficient and head coefficient. It is not possible to say what the functional relationship is at this stage, since it must be obtained by experiment on a particular prototype machine or model. In the case of a hydraulic machine, it is found that the Reynolds number is usually very high and therefore the viscous action of the fluid has very little effect on the power output of the machine and the power coefficient remains only a function of ψ and ϕ. To see how $\bar{P}$ could vary with ϕ and ψ, let us return to Fig. 1.2.

To determine the relationship between P, ψ and ϕ, the head across the machine can be fixed, as is usually the case in a hydroelectric installation. For a fixed value of inlet valve opening, the load on the machine is varied while the torque, speed and flow rate are measured. From these measurements, the power may be calculated, and $\bar{P}$ and ϕ plotted against ψ.

Typical dimensionless characteristic curves for a hydraulic turbine and pump are shown in Figs 1.3a and 1.3b, respectively. These curves are also the

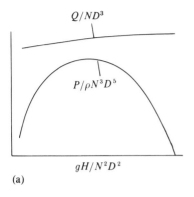
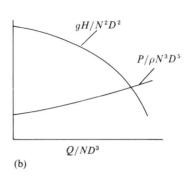

(a)
(b)

Figure 1.3 Performance characteristics of hydraulic machines drawn in terms of dimensionless groups: (a) hydraulic turbine; (b) hydraulic pump

characteristics of any other combination of P, N, Q and H for a given machine or for any other geometrically similar machine of different diameter. Since these groups are dimensionless, they may be divided or multiplied by themselves to form other dimensionless groups depending on the type of test being carried out, and it therefore follows that while in this particular case solutions for a, b and d were found in terms of c, e and f, other solutions could have been determined which give different dimensionless groups. Each set of groups taken together is correct, although they will of course be related by differently shaped curves.

For the turbine, the hydraulic efficiency is defined as

$$\eta = \frac{\text{Power delivered to runner}}{\text{Power available to runner}} \tag{1.7}$$

$$= P/\rho g Q H$$

Then substituting for P and rearranging gives

$$\eta = \bar{P}(ND^3/Q)(N^2D^2/gH)$$
$$= \bar{P}/\phi\psi \tag{1.8}$$

For a pump

$$\eta = \phi\psi/\bar{P} \tag{1.9}$$

1.3.2 Model Testing

Many hydraulic machines are so large that only a single unit might be required, as for example a hydraulic turbine in a hydroelectric installation producing many megawatts (MW) of power. Therefore, before the full-size machine is built, it is necessary to test it in model form to obtain as much information as possible about its characteristics. So that we may accurately transpose the results obtained from the model to the full-size machine, three criteria must be met. The first is that the model and prototype must be geometrically similar; that is, the ratio of all lengths between the model and prototype must be the same. The second requirement is that of kinematic similarity, where the velocities of the fluid particles at corresponding points in the model and prototype must be related through a fixed ratio. The third requirement is that of dynamic similarity, where the forces acting at corresponding points must be in a fixed ratio between model and prototype. For a geometrically similar model, dynamic similarity implies kinematic similarity.

In order to ensure the above criteria, the values of the dimensionless groups in Eq. (1.5) must remain the same for both the model and the prototype. Therefore if the curves shown in Fig. 1.3 had been obtained for a completely similar model, these same curves would apply to the full-size prototype machine. It can then be seen that these curves apply to any size machine of the same family at any head, flow rate or speed.

1.3.3 Compressible Flow Machines

Not all turbomachines use a liquid (hydraulic fluid) as their fluid medium. Gas turbines and axial flow compressors are used extensively in the jet engines of aircraft where the products of combustion and air respectively are the working fluids, while many diesel engines use centrifugal compressors for supercharging. To accommodate the compressibility of these types of fluids (gases), some new variables must be added to those already mentioned in the case of hydraulic machines, and changes must be made in some of the definitions used.

With compressible flow machines, the parameters of importance are the pressure and temperature increase of the gas in a compressor and the pressure and temperature decrease of the gas in the turbine plotted as a function of the mass flow rate of the gas. In Fig. 1.4, the $T–s$ charts for a compression and expansion process are shown.

In isentropic flow the outlet conditions of the gas are at 02s whereas the actual outlet conditions are at 02. The subscript 0 refers to total conditions and 1 and 2 refer to the inlet and outlet points of the gas respectively. The s refers to constant entropy.

Now the pressure at the outlet, p_{02}, can be written as a function of the following variables:

$$p_{02} = f(D, N, m, p_{01}, T_{01}, T_{02}, \rho_{01}, \rho_{02}, \mu) \qquad (1.10)$$

Here the pressure ratio p_{02}/p_{01} replaces the head H in the hydraulic machine, while the mass flow rate m (kg/s) replaces Q. However, by examining Eq. (1.10) we can see that, using the equation of state, the density may be written as

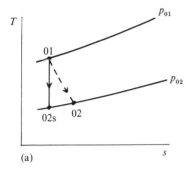 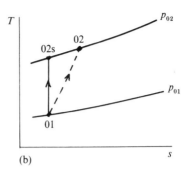

Figure 1.4 Compression and expansion in compressible flow machines: (a) turbine; (b) compressor

$\rho = p/RT$ and it therefore becomes superfluous since we already have T and p as variables, so deleting density, and combining R with T, the functional relationship can be written as

$$p_{02} = f(p_{01}, RT_{01}, RT_{02}, m, N, D, \mu)$$

and writing p_{02} as a product of the terms raised to powers,

$$p_{02} = \text{const}\,[(p_{01})^a (RT_{01})^b (RT_{02})^c (m)^d (N)^e (D)^f (\mu)^g] \tag{1.11}$$

Putting in the basic dimensions

$$(M/LT^2) = \text{const}\,[(M/LT^2)^a (L^2/T^2)^b (L^2/T^2)^c (M/T)^d (1/T)^e (L)^f (M/LT)^g]$$

Equating indices

$$
\begin{aligned}
\text{M} \qquad & 1 = a + d + g \\
\text{L} \qquad & -1 = -a + 2b + 2c + f - g \\
\text{T} \qquad & -2 = -2a - 2b - 2c - d - e - g
\end{aligned}
$$

and solving for a, b and f in terms of d, c, e and g we obtain

$$
\begin{aligned}
a &= 1 - d - g \\
b &= d/2 - c - e/2 + g/2 \\
f &= e - 2d - g
\end{aligned}
$$

Substitute for a, b and f in Eq. (1.11), then

$$
\begin{aligned}
p_{02} &= \text{const}\,[p_{01}^{1-d-g}(RT_{01})^{d/2-c-e/2+g/2}(RT_{02})^c m^d N^e D^{e-2d-g}\mu^g] \\
&= \text{const} \times p_{01}\{(RT_{02}/RT_{01})^c [m(RT_{01})^{1/2}/p_{01}D^2]^d [ND/(RT_{01})^{1/2}]^e \\
&\quad \times [\mu(RT_{01})^{1/2}/p_{01}D]^g\} \tag{1.12}
\end{aligned}
$$

Now if the last term in the brackets in Eq. (1.12) is multiplied top and bottom by $(RT_{01})^{1/2}$ and noting that p_{01}/RT_{01} equals ρ_{01}, then

$$\mu RT_{01}/p_{01}(RT_{01})^{1/2}D = \mu/(RT_{01})^{1/2}\rho_{01}D$$

But the units of $(RT_{01})^{1/2}$ are L/T, which is a velocity, and therefore the last term in brackets is expressible as a Reynolds number. Thus the functional relationship may be written as

$$p_{02}/p_{01} = f((RT_{02}/RT_{01}), (m(RT_{01})^{1/2}/p_{01}D^2), (ND/(RT_{01})^{1/2}), Re) \quad (1.13)$$

The exact form of the function (1.13) must be obtained by experimental measurements taken from model or prototype tests. For a particular machine using a particular fluid, or for a model using the same fluid as the prototype, R is a constant and may be eliminated. The Reynolds number is in most cases so high and the flow so turbulent that changes in this parameter over the usual operating range may be neglected. However, where large changes of density

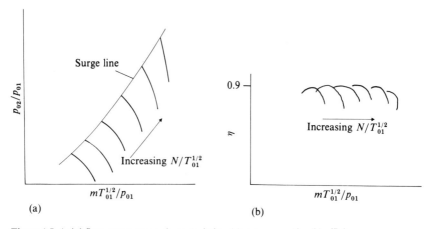

Figure 1.5 Axial flow compressor characteristics: (a) pressure ratio; (b) efficiency

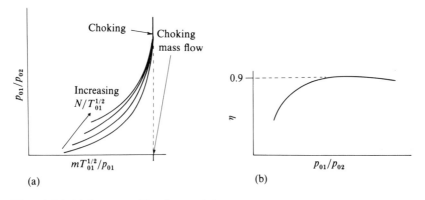

Figure 1.6 Axial flow gas turbine characteristics: (a) pressure ratio; (b) efficiency

take place, a significant reduction in Re can occur, and this must then be taken into account. For a particular constant-diameter machine, the diameter D may be ignored and therefore, in view of the above considerations, function (1.13) becomes

$$p_{02}/p_{01} = f((T_{02}/T_{01}), (mT_{01}^{1/2}/p_{01}), (N/T_{01}^{1/2})) \qquad (1.14)$$

where it should be noted that some of the terms are now no longer dimensionless. It is usual to plot p_{02}/p_{01} and T_{02}/T_{01} against the mass flow rate parameter $mT_{01}^{1/2}/p_{01}$ for different values of the speed parameter $N/T_{01}^{1/2}$ for a particular machine. But for a family of machines, the full dimensionless groups of Eq. (1.13) must be used if it is required to change the size of the machine or the gas contained. The term $ND/(RT_{01})^{1/2}$ can be interpreted as the Mach-number effect. This is because the impeller velocity $V \propto ND$ and the acoustic velocity $a_{01} \propto (RT_{01})^{1/2}$, while the Mach number $M = V/a_{01}$. Typical performance curves for an axial flow compressor and turbine are shown in Figs 1.5 and 1.6.

1.4 PROTOTYPE AND MODEL EFFICIENCY

Before leaving this introduction to the use of dimensionless groups, let us look at the relationship between the efficiency of the model and that of the prototype, assuming that the similarity laws are satisfied.

We wish to build a model of a prototype hydraulic turbine of efficiency η_p. Now from similarity laws, denoting the model and prototype by subscripts m and p respectively,

$$H_p/(N_pD_p)^2 = H_m/(N_mD_m)^2 \quad \text{or} \quad H_p/H_m = (N_p/N_m)^2(D_p/D_m)^2$$
$$Q_p/N_pD_p^3 = Q_m/N_mD_m^3 \quad \text{or} \quad Q_p/Q_m = (N_p/N_m)(D_p/D_m)^3$$
$$P_p/N_p^3D_p^5 = P_m/N_m^3D_m^5 \quad \text{or} \quad P_m/P_p = (N_m/N_p)^3(D_m/D_p)^5$$

Now

$$\text{Turbine efficiency} = \frac{\text{Power transferred from fluid}}{\text{Fluid power available}}$$

$$= P/\rho g Q H \qquad (1.15)$$

Therefore

$$\eta_m/\eta_p = (P_m/P_p)(Q_p/Q_m)(H_p/H_m) = 1$$

and the efficiencies of the model and prototype are the same providing the similarity laws are satisfied. In practice, the two are not the same due to scaling effects, such as relative surface roughness, slight Reynolds-number changes and Mach-number effects at higher blade speeds.

1.5 DIMENSIONLESS SPECIFIC SPEED

We have seen in Sec. 1.3 that the curves showing the functional relationship between dimensionless groups for a particular machine also apply to machines of the same family (similar design), providing the similarity laws are obeyed when changing to a smaller-diameter machine, at perhaps a different speed and head. It is therefore possible to obtain curves of many different types of machines, and to use these curves to select a machine design for a particular operating requirement. Typical curves that might be obtained for different types of hydraulic pumps are shown in Fig. 1.7, where it is seen that each machine type lies in a well-defined region of head and flow coefficients, it being possible in some cases to choose two or more impeller types for a specific flow coefficient. There are of course an infinite number of designs that could be produced, but for each design only one point exists on its characteristic curve where the efficiency is at a maximum. Thus for each design of pump unique values of ϕ and ψ exist at the maximum efficiency point. In the case of turbines, the unique values would be $\bar{P}$ and ϕ at maximum efficiency.

The specifications for a pump design are usually expressed in terms of a required head H, at a flow rate of Q and speed N, the speed being specified since motors are usually only available in fixed speed intervals. No mention has been made concerning the diameter or type of machine, both of which must be determined. For the best design point, constant values of ϕ_D and ψ_D will exist corresponding to the maximum efficiency point, or

$$\phi_D = Q/ND^3 \quad \text{and} \quad \psi_D = gH/N^2D^2$$

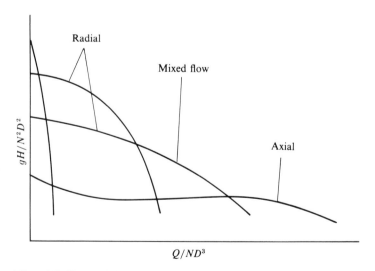

Figure 1.7 Characteristic curves for various pump designs

If the diameter is eliminated from these two equations, then

$$D = (gH/\psi_D)^{1/2}/N \qquad \text{and} \qquad \phi_D = QN^2(\psi_D/gH)^{3/2}$$

or

$$NQ^{1/2}/(gH)^{3/4} = (\phi_D^{1/2}/\psi_D^{3/4}) = \text{const} = N_s \qquad (1.16)$$

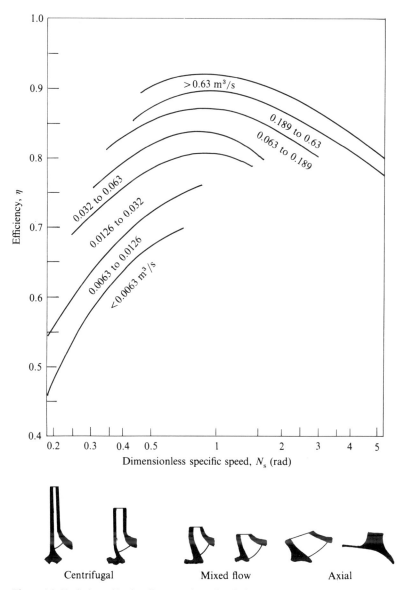

Figure 1.8 Variation of hydraulic pump impeller design

N_s is known as the dimensionless specific speed, the units being revolution or radians depending on the units of N, and must not be confused with specific speed.

Since D was eliminated at the maximum efficiency point, the dimensionless specific speed acts as a design parameter, which indicates the type of machine that should be used for the given N, H and Q. Equation (1.16) shows that a pump with a high N_s will have a low head and high flow rate, and implies an axial flow pump with a large swallowing capacity. A low N_s implies a high head and low flow rate, and a centrifugal type of pump. Figure 1.8 shows the variation of N_s with pump impeller type, and indicates the optimum efficiencies to be expected.

In practice, N_s is often expressed as $NQ^{1/2}/H^{3/4}$, the g being dropped since it is a constant, and the resulting value of N_s will therefore be different. It may also be found that consistent sets of units are not always used for N, Q and H, so that when a value of N_s is expressed, it should be ensured that the definition being used is known. In this text the SI system will be used and N_s will therefore be dimensionless.

However, as a point of reference, conversion factors are listed in Table 1.2 so that the reader may calculate the dimensionless specific speed from specific speeds using Q, N and H in other units. The fluid contained is water and, where quoted, gpm are US gallons per minute, ft is foot, cfs are cubic feet per second, and hp is horsepower.

Terms that are often used in hydraulic flow machines are those of unit head, unit speed, unit power and unit quantity. They arose from the need to be able to compare hydraulic machines tested under a set of standard conditions. In turbine work, the speed, power output and flow rate are determined for a turbine operating under an assumed unit head of say 1 m or 1 ft, its efficiency remaining constant. For instance, consider a turbine tested under a head H_1 and speed N_1 rpm. Then from Eq. (1.6), for any other speed and head,

$$H_1/N_1^2 = H_2/N_2^2$$

or

$$N_2 = N_1(H_2/H_1)^{1/2}$$

Table 1.2 Conversion factors for specific speed

Specific speed	Dimensionless specific speed, N_s (rad)
$SP_1 = \text{rpm}(\text{cfs})^{1/2}/\text{ft}^{3/4}$	$N_s = SP_1/129$
$SP_2 = \text{rpm}(\text{m}^3/\text{s})^{1/2}/\text{m}^{3/4}$	$N_s = SP_2/53$
$SP_3 = \text{rpm}(\text{gpm})^{1/2}/\text{ft}^{3/4}$	$N_s = SP_3/2730$
$SP_4 = \text{rpm}(\text{hp})^{1/2}/\text{ft}^{5/4}$	$N_s = SP_4/42$
$SP_5 = \text{rpm}(\text{metric hp})^{1/2}/\text{m}^{5/4}$	$N_s = SP_5/187$

Putting $H_2 = 1$ (unit head) then

$$N_2 = N_1/H_1^{1/2} = N_u \qquad (1.17)$$

and this is the unit speed of the turbine. Unit quantities for Q and P may be similarly obtained to give

$$Q_u = Q/H^{1/2} \qquad (1.18)$$

and

$$P_u = P/H^{3/2} \qquad (1.19)$$

For a turbine, the dimensionless specific speed is found by a procedure similar to that for pumps except that D is eliminated from P and ψ to yield what is often referred to as the power specific speed, N_{sp}, where

$$N_{sp} = NP^{1/2}/\rho^{1/2}(gH)^{5/4} \qquad (1.20)$$

Figure 1.9 shows typical hydraulic turbine runner shapes for different specific speeds along with their optimum or design efficiencies.

A wide range of rotor designs from low to high values of specific speed for both hydraulic and compressible flow machines are shown in Figs 1.10 and 1.11, where it will be noted that low-specific-speed machines have large diameters and high-specific-speed machines have small diameters. In general,

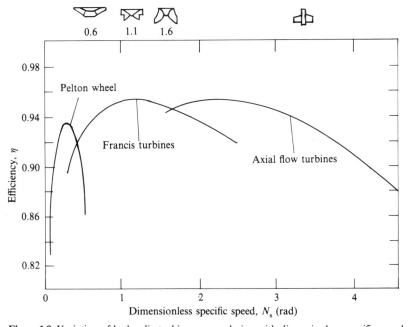

Figure 1.9 Variation of hydraulic turbine runner design with dimensionless specific speed

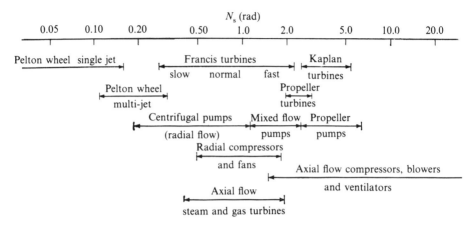

Figure 1.10 Correlation of rotor designs with dimensionless specific speed (*courtesy of Escher Wyss Ltd*)

the smaller the diameter the lower will be the cost of the machine, and therefore the design usually aims for the highest possible specific speed.

1.6 BASIC LAWS AND EQUATIONS

The basic laws of thermodynamics and fluid mechanics are used in turbomachines although they are usually arranged into a more convenient form. All or some may be used under any set of circumstances and each will be briefly dealt with in turn.

1.6.1 Continuity

For steady flow through the control volume, the mass flow rate m remains constant. Referring to Fig. 1.12,

$$m = \rho_1 C_1 A_1 = \rho_2 C_2 A_2 \tag{1.21}$$

where the velocity vectors C_1 and C_2 are perpendicular to the cross-sectional areas of flow A_1 and A_2. In compressible flow machines the mass flow (kg/s) is used almost exclusively while in hydraulic machines the volume flow rate Q (m^3/s) is preferred.

1.6.2 Steady Flow Energy Equation (First Law of Thermodynamics)

For steady flow through a system control volume, where the heat transfer rate to the system from the surroundings is Q and the work done by the system on

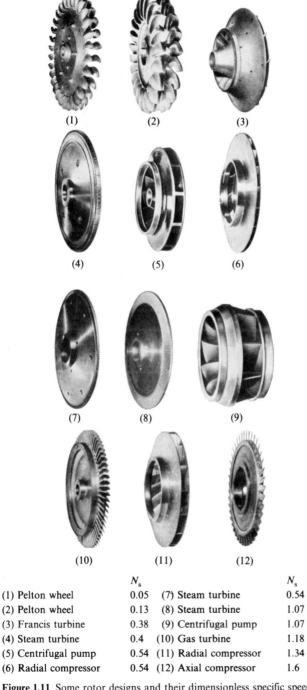

		N_s			N_s
(1)	Pelton wheel	0.05	(7)	Steam turbine	0.54
(2)	Pelton wheel	0.13	(8)	Steam turbine	1.07
(3)	Francis turbine	0.38	(9)	Centrifugal pump	1.07
(4)	Steam turbine	0.4	(10)	Gas turbine	1.18
(5)	Centrifugal pump	0.54	(11)	Radial compressor	1.34
(6)	Radial compressor	0.54	(12)	Axial compressor	1.6

Figure 1.11 Some rotor designs and their dimensionless specific speeds

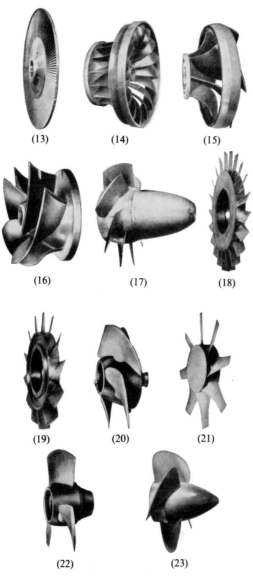

	N_s		N_s
(13) Steam turbine	1.6	(19) Axial compressor	3.21
(14) Francis turbine	1.72	(20) Propeller pump	3.21
(15) Francis turbine	2.14	(21) Axial blower	4.82
(16) Mixed flow pump	2.14	(22) Propeller pump	5.36
(17) Kaplan turbine	2.41	(23) Kaplan turbine	5.36
(18) Axial compressor	2.41		

Figure 1.11 (*contd*)

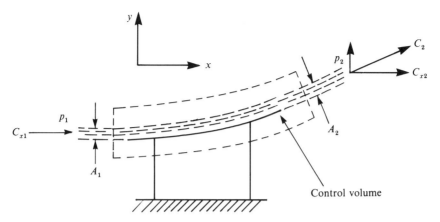

Figure 1.12 Control volume for linear momentum

the surroundings is W, then

$$Q - W = m[(p_2/\rho_2 - p_1/\rho_1) + (C_2^2 - C_1^2)/2 + g(Z_2 - Z_1) + (u_2 - u_1)] \quad (1.22)$$

where $p/\rho = $ pressure energy per unit mass (J/kg), $C^2/2 = $ kinetic energy per unit mass (J/kg), $u = $ internal energy of the fluid per unit mass (J/kg), gZ = potential energy per unit mass (J/kg), $m = $ mass flow rate (kg/s), $W = $ work done *on* surroundings ($+$ve) (W) and $Q = $ heat transfer *to* system ($+$ve) (W).

In words, Eq. (1.22) states that in steady flow through any region:

Heat added to	Shaft work done		Increase in		Increase in
fluid per unit	$-$ by the fluid per	$=$	pressure energy	$+$	kinetic energy
mass	unit mass		per unit mass		per unit mass

	Increase in		Increase in
	$+$ potential energy	$+$	internal energy
	per unit mass		per unit mass

The steady flow energy equation applies to liquids, gases and vapours as well as to real fluids having no viscosity. It may be simplified in many cases because many of the terms are zero or cancel with others, and this will be shown in the relevant sections.

1.6.3 Newton's Second Law of Motion

This law states that the sum of all the forces acting on a control volume in a particular direction is equal to the rate of change of momentum of the fluid across the control volume. With reference to Fig. 1.12, Newton's second law may be written as

$$\sum F_x = m(C_{x2} - C_{x1}) \quad (1.23)$$

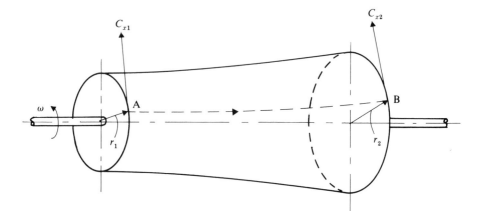

Figure 1.13 Control volume for angular momentum

Equation (1.23) applies for linear momentum. However, turbomachines have impellers that rotate, and the power output is expressed as the product of torque and angular velocity, and therefore angular momentum is the more useful parameter.

Figure 1.13 shows the movement of a fluid particle from a point A to a point B while at the same time moving from a radius r_1 to r_2. If the tangential velocities of the fluid are C_{x1} and C_{x2} respectively, then the sum of all the torques acting on the system is equal to the rate of change of angular momentum,

$$\sum T = m(r_2 C_{x2} - r_1 C_{x1})$$

If the machine revolves with angular velocity ω then the power is

$$\sum T\omega = m(U_2 C_{x2} - U_1 C_{x1})$$

For a turbine

$$W = m(U_1 C_{x1} - U_2 C_{x2}) > 0 \qquad (1.24)$$

and is known as Euler's turbine equation.

For a pump

$$W = m(U_2 C_{x2} - U_1 C_{x1}) > 0 \qquad (1.25)$$

which is Euler's pump equation.

1.6.4 Entropy (Second Law of Thermodynamics)

This law states that, for a fluid undergoing a reversible adiabatic process, the entropy change is zero, while for the same fluid undergoing an adiabatic or other process, the entropy increases from inlet to outlet. It is this fact that causes the power developed by a turbine to be less than the ideal isentropic power developed and why the work input to a pump is greater than the isentropic or ideal work input (Fig. 1.4). In theory the entropy change might also be zero for an adiabatic process but it is impossible in practice. For a reversible process the second law is expressed as

$$dq/T = \Delta s \qquad (1.26)$$

where dq = heat transfer per unit mass (J/kg), T = absolute temperature at which heat transfer occurs (K) and Δs = entropy change (J/kg K).

In the absence of motion, gravity and any other effects, Eq. (1.22) has no potential or kinetic energy terms, and so

$$Q - W = m(u_2 - u_1) \qquad \text{or} \qquad dq - dw = du$$

where the units are J/kg. Substituting for dq and rearranging,

$$\Delta s = dq/T = (du + dw)/T$$

Putting $dw = p\,dv$, where dv is an infinitesimal specific volume change, then

$$T\,ds = du + p\,dv \qquad (1.27)$$

Defining specific enthalpy as $h = u + pv$ and substituting for du in Eq. (1.27),

$$T\,ds = dh - v\,dp \qquad (1.28)$$

and this equation is used extensively in the study of compressible flow machines.

In the following chapters, use will be made of the concepts discussed in this introduction. This chapter should have acted as a reminder of the many separate concepts learned in thermodynamics and fluid mechanics, and has shown how these two separate subjects combine to form the subject of turbomachinery.

EXERCISES

1.1 A radial flow hydraulic turbine is required to be designed to produce 30 MW under a head of 14 m at a speed of 95 rpm. A geometrically similar model with an output of 40 kW and a head of 5 m is to be tested under dynamically similar conditions. At what speed must the model be run, what is the required impeller diameter ratio between the model and prototype and what is the volume flow rate through the model if its efficiency can be assumed to be 90 per cent?

1.2 The performance curves of a centrifugal pump are shown in Fig. 1.14. The impeller diameter is 127 mm and the pump delivers 2.83 l/s at a speed of 2000 rpm. If a 102 mm diameter impeller is fitted and the pump runs at a speed of 2200 rpm, what is the volume flow rate? Determine also the new pump head.

1.3 An axial flow compressor is designed to run at 4500 rpm when ambient atmospheric conditions are 101.3 kPa and 15°C. On the day when the performance characteristic is obtained, the atmospheric temperature is 25°C. What is the correct speed at which the compressor must run?

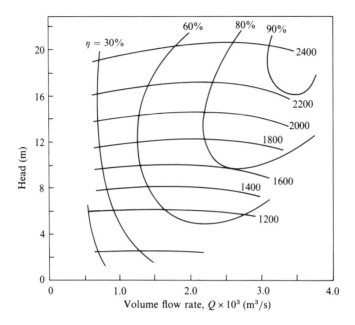

Figure 1.14

If an entry pressure of 60 kPa is obtained at the point where the normal ambient condition mass flow would be 65 kg/s, calculate the mass flow obtained in the test.

1.4 Specifications for an axial flow coolant pump for one loop of a pressurized water nuclear reactor are:

Head 85 m
Flow rate 20 000 m³/h
Speed 1490 rpm
Diameter 1200 mm
Water density 714 kg/m³
Power 4 MW (electrical)

The manufacturer plans to build a model. Test conditions limit the available electric power to 500 kW and flow to 0.5 m³/s of cold water. If the model and prototype efficiencies are assumed equal, find the head, speed and scale ratio of the model. Calculate the dimensionless specific speed of the prototype and confirm that it is identical with the model.

1.5 A pump with an available driven speed of 800 rpm is required to overcome a 1.83 m head while pumping 0.2 m³/s. What type of pump is required and what power is required?

1.6 A reservoir has a head of 40 m and a channel leading from the reservoir permits a flow rate of 34 m³/s. If the rotational speed of the rotor is 150 rpm, what is the most suitable type of turbine to use?

1.7 A large centrifugal pump contains liquid whose kinematic viscosity may vary between 3 and 6 times that of water. The dimensionless specific speed of the pump is 0.183 rev and it is to discharge 2 m³/s of liquid against a total head of 15 m. Determine the range of speeds and test heads for a one-quarter scale model investigation of the full size pump if the model uses water.

1.8 In a projected low-head hydroelectric scheme, 10 000 ft³/s of water are available under a head of 12 ft. Alternative schemes to use Francis turbines having a specific speed of 105 rpm or Kaplan turbines with a specific speed of 180 rpm are investigated. The normal running speed is to be 50 rpm in both schemes. Determine the dimensionless specific speeds and compare the two proposals insofar as the number of machines are concerned, and estimate the power to be developed by each machine. The units in either installation are to be of equal power and the efficiency of each type may be assumed to be 0.9.

1.9 A customer approaches a salesman with a particular pump requirement and is quoted for an axial flow pump of rotor diameter 152.4 mm. Running at a speed of 980 rpm, the machine is said to deliver $0.283\,\text{m}^3/\text{s}$ of water against a head of 9.1 m at an efficiency of 85 per cent. Are the claims of the salesman realistic?

1.10 A Francis turbine runs at 180 rpm under a head of 146 m with an efficiency of 93.5 per cent. Estimate the power output of the installation.

SOLUTIONS

Exercise 1.1 Equating head, flow and power coefficients for the model and prototype and noting that the density of the fluid remains the same, then, if subscript 1 refers to the prototype and subscript 2 to the model,

$$\frac{P_1}{\rho_1 N_1^3 D_1^5} = \frac{P_2}{\rho_2 N_2^3 D_2^5} \qquad \text{where } \rho_1 = \rho_2$$

Then

$$\frac{D_2}{D_1} = \left(\frac{P_2}{P_1}\right)^{1/5}\left(\frac{N_1}{N_2}\right)^{3/5}$$

$$= \left(\frac{0.04}{30}\right)^{1/5}\left(\frac{N_1}{N_2}\right)^{3/5}$$

$$= 0.266\left(\frac{N_1}{N_2}\right)^{3/5}$$

Also

$$\frac{gH_1}{(N_1 D_1)^2} = \frac{gH_2}{(N_2 D_2)^2}$$

Then

$$\frac{D_2}{D_1} = \left(\frac{H_2}{H_1}\right)^{1/2}\left(\frac{N_1}{N_2}\right) = \left(\frac{5}{14}\right)^{1/2}\left(\frac{N_1}{N_2}\right)$$

Therefore equating the diameter ratios

$$0.266\left(\frac{N_1}{N_2}\right)^{3/5} = \left(\frac{5}{14}\right)^{1/2}\left(\frac{N_1}{N_2}\right)$$

or

$$\left(\frac{N_2}{N_1}\right)^{2/5} = 2.25$$

whence

$$N_2 = 2.25^{5/2} \times 95$$

$$\underline{\text{Model speed} = 721.4 \, \text{rpm}}$$

$$\frac{D_2}{D_1} = 0.266 \left(\frac{95}{721.4} \right)^{3/5}$$

$$\underline{\text{Model scale ratio} = 0.079}$$

$$\text{Model efficiency} = \frac{\text{Power output}}{\text{Water power input}}$$

$$0.9 = \frac{40 \times 10^3}{\rho g Q H}$$

$$Q = \frac{40 \times 10^3}{0.9 \times 10^3 \times 9.81 \times 5}$$

$$\underline{\text{Model volume flow} = 0.906 \, \text{m}^3/\text{s}}$$

Exercise 1.2 Assuming dynamic similarity exists between the first and second sized pumps, we equate the flow coefficients. Thus

$$\frac{Q_1}{N_1 D_1^3} = \frac{Q_2}{N_2 D_2^3}$$

$$\frac{2.83}{2000 \times 127^3} = \frac{Q_2}{2200 \times 102^3}$$

Solving we get

$$Q_2 = 1.61 \, \text{l/s}$$

From Fig. 1.14 at $Q_1 = 2.83 \, \text{l/s} \, (2.83 \times 10^{-3} \, \text{m}^3/\text{s})$ and 2000 rpm the head H_1 is 14 m and equating head coefficients for both cases gives

$$\frac{g H_1}{N_1^2 D_1^2} = \frac{g H_2}{N_2^2 D_2^2}$$

and substituting

$$\frac{9.81 \times 14}{(2000 \times 127)^2} = \frac{9.81 \times H_2}{(2200 \times 102)^2}$$

Solving we get

$$H_2 = 10.9 \, \text{m of water}$$

Exercise 1.3 The dimensionless groups of Eq. (1.13) may be used here but since the same machine is being considered in both cases the gas constant R

and diameter D are dropped to yield Eq. (1.14). Considering first the speed parameter,

$$\frac{N_1}{\sqrt{T_{01}}} = \frac{N_2}{\sqrt{T_{02}}}$$

$$N_2 = 4500 \left(\frac{273 + 25}{273 + 15}\right)^{1/2}$$

Correct speed = 4577 rpm

Considering now the mass flow parameter,

$$\frac{m_1 \sqrt{T_{01}}}{p_{01}} = \frac{m_2 \sqrt{T_{02}}}{p_{02}}$$

$$m_2 = 65 \times \left(\frac{60}{101.3}\right)\left(\frac{288}{298}\right)^{1/2}$$

Mass flow obtained = 37.85 kg/s

Exercise 1.4 Using Eq. (1.5), equate the head power and flow coefficients for the model and prototype. Then

$$\frac{Q_1}{Q_2} = \left(\frac{N_1}{N_2}\right)\left(\frac{D_1}{D_2}\right)^3$$

or

$$\frac{N_1}{N_2} = \left(\frac{20\,000}{0.5 \times 3600}\right)\left(\frac{D_2}{D_1}\right)^3$$

$$= 11.11\left(\frac{D_2}{D_1}\right)^3$$

Also

$$\frac{P_1}{P_2} = \left(\frac{N_1}{N_2}\right)^3\left(\frac{D_1}{D_2}\right)^5\left(\frac{\rho_1}{\rho_2}\right)$$

Substitute for (N_1/N_2); then

$$\frac{4}{0.5} = (11.11)^3\left(\frac{D_2}{D_1}\right)^9\left(\frac{D_1}{D_2}\right)^5\left(\frac{714}{1000}\right)$$

$$\left(\frac{D_2}{D_1}\right)^4 = \frac{8}{(11.11)^3 \times 0.714}$$

Scale ratio $D_2/D_1 = 0.3$

Then

$$N_1/N_2 = 11.11 \times (0.3)^3$$

Speed ratio $N_2/N_1 = 3.3$

Also

$$\frac{H_2}{H_1} = \left(\frac{N_1}{N_2}\right)^2 \left(\frac{D_1}{D_2}\right)^2$$

$$= \left(\frac{1}{3.3}\right)^2 \left(\frac{1}{0.3}\right)^2$$

Head ratio $H_2/H_1 = 1.0$

From Eq. (1.16) the dimensionless specific speed is given by

$$N_s = \frac{NQ^{1/2}}{(gH)^{3/4}}$$

For the prototype

$$N_{s1} = \frac{2\pi \times 1490}{60} \times \left(\frac{20\,000}{3600}\right)^{1/2} \left(\frac{1}{9.81}\right)^{3/4} \left(\frac{1}{85}\right)^{3/4}$$

$$= 2.37 \text{ rad}$$

$$N_{s2} = 2\pi \times \frac{1490}{60} \times 3.3 \times \frac{(0.5)^{1/2}}{(9.81 \times 85)^{3/4}}$$

$$= 2.35 \text{ rad}$$

Therefore taking rounding errors into account the dimensionless specific speeds of both model and prototype are the same.

Exercise 1.5 From Eq. (1.16)

$$N_s = \frac{NQ^{1/2}}{(gH)^{3/4}}$$

$$= \frac{800}{60} \times \frac{(0.2)^{1/2}}{(9.81 \times 1.83)^{3/4}}$$

$$= 0.683 \text{ rev } (4.29 \text{ rad})$$

For the given flow rate Fig. 1.8 shows that a propeller or axial flow pump is required and that an efficiency of about 80 per cent can be expected. Therefore the power required is

$$P = \rho g Q H$$

$$= 1000 \times 9.81 \times 0.2 \times 1.83$$

$$= 3.59 \text{ kW}$$

This is the power delivered to the water and to get the power that must be supplied to the shaft we divide by the efficiency,

$$\text{Shaft power required} = 3.59/0.80$$
$$\underline{\text{Shaft power} = 4.49\,\text{kW}}$$

Exercise 1.6 We have

$$\text{Turbine power} = \rho g Q H$$
$$= 1000 \times 9.81 \times 34 \times 40$$
$$= 13.3\,\text{MW}$$

Power specific speed is given by Eq. (1.20)

$$N_{sp} = \frac{NP^{1/2}}{\rho^{1/2}(gH)^{5/4}}$$

$$= \frac{150 \times (13.3 \times 10^6)^{1/2}}{60 \times (1000)^{1/2} \times (9.81 \times 40)^{5/4}}$$

$$\underline{= 0.165\,\text{rev}\,(1.037\,\text{rad})}$$

From Fig. 1.10 it is seen that the Francis turbine would be the most suitable choice for this application.

Exercise 1.7 Since the viscosity of the liquids used in the model and prototype vary significantly, equality of Reynolds number in Eq. (1.5) must apply for dynamic similarity. Let subscripts 1 and 2 apply to the prototype and model respectively.

Equating Reynolds number

$$\frac{N_1 D_1^2}{v_1} = \frac{N_2 D_2^2}{v_2}$$

$$\frac{N_2}{N_1} = \frac{v_2}{v_1}\left(\frac{D_1}{D_2}\right)^2$$

For the liquid with viscosity three times that of water

$$\frac{N_2}{N_1} = \frac{4^2}{3} = 5.333$$

Equating flow coefficients

$$\frac{Q_1}{N_1 D_1^3} = \frac{Q_2}{N_2 D_2^3}$$

$$\frac{Q_2}{Q_1} = \frac{N_2}{N_1}\left(\frac{D_2}{D_1}\right)^3$$

$$= \frac{5.333}{4^3} = 0.0833$$

Equating head coefficients

$$\frac{H_1}{(N_1 D_1)^2} = \frac{H_2}{(N_2 D_2)^2}$$

$$\frac{H_2}{H_1} = \left(\frac{N_2}{N_1}\right)^2 \left(\frac{D_2}{D_1}\right)^2$$

$$= \left(\frac{5.33}{4}\right)^2 = 1.776$$

From Eq. (1.16)

$$N_{s1} = \frac{N_1 Q_1^{1/2}}{(gH_1)^{3/4}}$$

$$N_1 = \frac{0.183(9.81 \times 15)^{3/4}}{2^{1/2}}$$

$$= 5.47 \, \text{rev/s}$$

$$N_2 = 5.47 \times 5.33$$

$$\underline{\text{Model speed} = 29.16 \, \text{rev/s}}$$

$$Q_2 = 2 \times 0.0833$$

$$\underline{\text{Model flow rate} = 0.166 \, \text{m}^3}$$

$$H_2 = 15 \times 1.776$$

$$\underline{\text{Model head} = 26.67 \, \text{m}}$$

Similarly for the case when the prototype viscosity is six times that of water

$$\underline{N_2 = 14.58 \, \text{rev/s}}$$

$$\underline{H_2 = 6.67 \, \text{m}}$$

For one-quarter scale model

$$\underline{14.58 < \text{model speed} < 29.16 \, \text{rev/s}}$$

$$\underline{6.67 \, \text{m} < \text{model head} < 26.67 \, \text{m}}$$

Exercise 1.8 The dimensionless specific speed is obtained from the conversion factors for specific speed given in Sec. 1.5. In this case for the non-SI units used

$$N_{sp} = SP/42$$

Dimensionless specific speed of Francis turbine

$$= 105/42$$

$$= 2.5 \, \text{rad}$$

Dimensionless specific speed of Kaplan turbine

$$= 180/42$$

$$= 4.3 \, \text{rad}$$

These values may be checked against those values in Fig. 1.10.
    Converting to SI units

$$10\,000 \, \text{ft}^3/\text{s} = 283.17 \, \text{m}^3/\text{s}$$

$$12 \, \text{ft} = 3.66 \, \text{m}$$

$$\text{Turbine efficiency} = \frac{\text{Power delivered}}{\text{Power available}}$$

$$0.9 = \frac{P}{\rho g Q H}$$

$$P = 0.9 \times 1000 \times 283.17 \times 3.66 \times 9.81$$

$$= 9150 \, \text{kW}$$

This is the total power delivered by all the turbines.
    Now

$$N_s = \frac{N P^{1/2}}{\rho^{1/2}(gH)^{5/4}} \qquad \text{where } N \text{ is in rev/s}$$

For the Francis turbine

$$2.5 = \frac{50 \times 2 \times \pi \times P^{1/2}}{60 \times (1000)^{1/2}(9.81 \times 3.66)^{5/4}}$$

whence

$$P = 1761 \, \text{kW}$$

$$\text{Number of Francis turbines} = \frac{\text{Total power required}}{\text{Power per machine}}$$

$$= \frac{9150}{1761}$$

$$= 5.19 \, (\text{say 6 machines})$$

For the Kaplan turbine

$$4.3 = \frac{50 \times 2 \times \pi \times P^{1/2}}{60 \times (1000)^{1/2}(9.81 \times 3.66)^{5/4}}$$

whence

$$P = 5209.7 \, \text{kW}$$

$$\text{Number of Kaplan turbines} = \frac{9150}{5209.7}$$

$$= 1.76 \, (\text{say } 2)$$

Exercise 1.9 From Eq. (1.16) for dimensionless specific speed

$$N_s = \frac{NQ^{1/2}}{(gH)^{3/4}} \, \text{rev}$$

$$= \frac{2 \times \pi \times 980 \times 0.283^{1/2}}{60 \times (9.81 \times 9.1)^{3/4}} \, \text{rad}$$

$$= 1.88 \, \text{rad}$$

Referring to Fig. 1.10 it is seen that axial flow pumps only begin at a dimensionless specific speed of approximately 2.0 rad. It is therefore unlikely that the salesman's claims are realistic.

A suitable pump would be of the mixed flow type which gives the stated efficiency at the required flow rate and calculated dimensionless specific speed.

Exercise 1.10 Using Fig. 1.9 the Francis turbine has an efficiency of 93.5 per cent at a dimensionless specific speed of 2.0 rad. From Eq. (1.20) the dimensionless power specific speed is

$$N_{sp} = \frac{NP^{1/2}}{\rho^{1/2}(gH)^{5/4}}$$

whence

$$P^{1/2} = \frac{2.0 \times (1000)^{1/2} \times (9.81 \times 146)^{5/4} \times 60}{180 \times 2 \times \pi}$$

$$= 29\,563$$

and

$$P = 874 \, \text{MW}$$

HYDRAULIC PUMPS

2.1 INTRODUCTION

The term 'hydraulics' is defined as the science of the conveyance of liquids through pipes. Most of the theory applicable to hydraulic pumps has been derived using water as the fluid medium but this by no means precludes the use of other liquids. Two types of pumps commonly used are centrifugal and axial flow types, so named because of the general nature of the fluid flow through the impeller. Both work on the principle that the energy of the liquid is increased by imparting tangential acceleration to it as it flows through the pump. This energy is supplied by the impeller, which in turn is driven by an electric motor or some other drive. In order to impart tangential acceleration to the liquid, rows of curved vanes or blades move transversely through it and the liquid is pushed sideways as it moves over the vanes as well as retaining its original forward component of velocity. Figure 1.1 showed typical centrifugal and axial flow pump impellers, while between these two extremes lie mixed flow pumps, which are a combination of centrifugal and axial flow pumps, part of the liquid flow in the impeller being axial and part radial.

The centrifugal and axial flow pumps will be dealt with in turn in the following sections. However, before considering the operation of each type in detail, we will look at a general pumping system, which is common to both types. This is shown in Fig. 2.1 where a pump (either axial or centrifugal) pumps liquid from a low to a high reservoir.

At any point in the system, the pressure, elevation and velocity can be expressed in terms of a total head measured from a datum line. For the lower reservoir the total head at the free surface is H_A and is equal to the elevation of

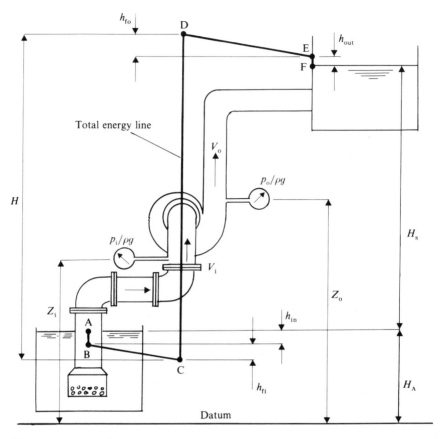

Figure 2.1 Diagram of a pumping system

the free surface above the datum line since the velocity and static gauge pressure at A are zero. The liquid enters the intake pipe causing the head loss h_{in}, with the result that the total head line drops to point B. As the fluid flows from the intake to the inlet flange of the pump at elevation Z_i, the total head drops further to the point C due to pipe friction and other losses h_{fi}. The fluid enters the pump and energy is imparted to it, which raises the total head to point D at the pump outlet. Flowing from the pump outlet to the upper reservoir, friction and other losses account for a total head loss h_{fo} down to point E, where an exit loss h_{out} occurs when the liquid enters the upper reservoir, bringing the total head at the upper reservoir to point F at the free surface.

If the pump total inlet and outlet heads are measured at the inlet and outlet flanges respectively, which is usually the case for a standard pump test,

then

$$\text{Pump total inlet head} = p_i/\rho g + V_i^2/2g + Z_i$$

$$\text{Pump total outlet head} = p_o/\rho g + V_o^2/2g + Z_o$$

$$\text{Total head developed by pump} = [(p_o - p_i)/\rho g] + [(V_o^2 - V_i^2)/2g] + (Z_o - Z_i)$$

$$= H \qquad (2.1)$$

This is the head that would be used in Eq. (1.12) for determining the type of pump that should be selected, and the term 'manometric head' is often used. The static head H_s is the vertical distance between the two levels in the reservoirs and from Fig. 2.1 it can be seen that for the pipeline

$$H = H_s + \sum \text{losses}$$

$$= H_s + h_{fi} + h_{fo} + h_{in} + h_{out}$$

It is worth noting here that, for the same size inlet and outlet diameters, V_o and V_i are the same, and in practice $(Z_o - Z_i)$ is so small in comparison to $(p_o - p_i)/\rho g$ that it is ignored. It is therefore not surprising to find that the static pressure head across the pump is often used to describe the total head developed by the pump.

2.2 CENTRIFUGAL PUMPS

Figure 2.2 shows the three important parts of a centrifugal pump: (1) the impeller, (2) the volute casing and (3) the diffuser ring.

The diffuser is optional and may or may not be present in a particular design depending upon the size and cost of the pump. The impeller is a rotating solid disc with curved blades standing out vertically from the face of the disc. The tips of the blades are sometimes covered by another flat disc to give shrouded blades; otherwise the blade tips are left open and the casing of the pump itself forms the solid outer wall of the blade passages. The advantage of the shrouded blade is that flow is prevented from leaking across blade tips from one passage to another. As the impeller rotates, the fluid that is drawn into the blade passages at the impeller inlet or eye is accelerated as it is forced radially outwards. In this way, the static pressure at the outer radius is much higher than at the eye inlet radius.

The fluid has a very high velocity at the outer radius of the impeller, and, to recover this kinetic energy by changing it into pressure energy, diffuser blades mounted on a diffuser ring may be used. The stationary blade passages so formed have an increasing cross-sectional area as the fluid moves through them, the kinetic energy of the fluid being reduced while the pressure energy is further increased. Vaneless diffuser passages may also be utilized.

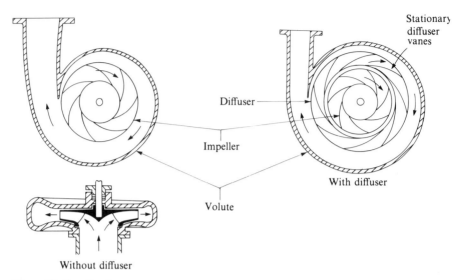

Figure 2.2 Centrifugal pump components

Finally, the fluid moves from the diffuser blades into the volute casing, which collects it and conveys it to the pump outlet. Sometimes only the volute casing exists without the diffuser; however, some pressure recovery will take place in the volute casing alone.

In dealing with the theory of hydraulic pumps, a number of assumptions will be made. At any point within the blade passages the fluid velocity will in general have three components, one each in the axial, radial and angular directions as indicated in Fig. 2.3.

The velocity may then be written as a function of the three components

$$V = f(r, \theta, z)$$

However, we will assume that the following hold:

1. There are an infinite number of blades so closely spaced that $\partial V / \partial \theta = 0$. That is, there is no flow in the blade passage in the tangential direction and $V_\theta = 0$.
2. The impeller blades are infinitely thin, thus allowing the pressure difference across them, which produces torque, to be replaced by tangential forces that act on the fluid.
3. The velocity variation across the width or depth of the impeller is zero and hence $\partial V / \partial z = 0$.
4. The analysis will be confined to conditions at the impeller inlet and outlet, and to the angular momentum change between these two stations. No account is taken of the condition of the fluid between these two stations.

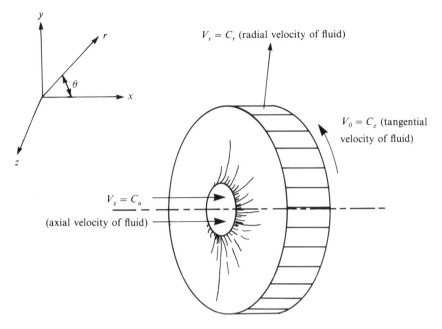

$V_r = C_r$ (radial velocity of fluid)

$V_\theta = C_x$ (tangential velocity of fluid)

$V_z = C_a$

(axial velocity of fluid)

Figure 2.3 Cylindrical coordinates for a centrifugal pump

5. It is assumed that at inlet the fluid is moving radially after entering the eye of the pump.

Assumptions 1 and 2 mean that the velocity is a function of the radius only, $V = f(r)$, and now with these assumptions the velocity vectors at inlet and outlet of the impeller can be drawn and the theoretical energy transfer determined.

Figure 2.4 shows a centrifugal pump impeller with the velocity triangles drawn at inlet and outlet. The blades are curved between the inlet radius r_1 and outlet radius r_2, a particle of fluid moving along the broken curve shown. β_1 is the angle subtended by the blade at inlet, measured from the tangent to the inlet radius, while β_2 is the blade angle measured from the tangent at outlet. The fluid enters the blade passages with an absolute velocity C_1 and at an angle α_1 to the impeller inlet tangential velocity vector U_1, where $U_1 = \omega r_1$, ω being the angular velocity of the impeller. The resultant relative velocity of flow into the blade passage is W_1 at an angle β_1' to the tangent at inlet. Similarly at outlet the relative velocity vector is W_2 at angle β_2' from the tangent to the blade. By subtracting the impeller outlet tangential velocity vector U_2, the absolute velocity vector C_2 is obtained, this being set at angle α_2 from the tangent to the blade. It is seen that the blade angles at inlet and outlet do not equal the relative flow angles at inlet and outlet. This is for a general case, and unless otherwise stated (see slip factor, Sec. 2.3), it will be assumed

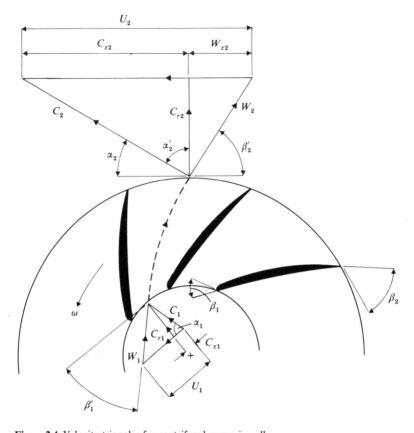

Figure 2.4 Velocity triangles for centrifugal pump impeller

that the inlet and outlet blade angles are equal to their corresponding relative flow angles.

From Euler's pump equation (Eq. (1.25)), the work done per second on the fluid per unit weight of fluid flowing is

$$E = W/mg = (U_2 C_{x2} - U_1 C_{x1})/g \quad \text{(J/s per N/s or m)} \tag{2.2}$$

where C_x is the component of absolute velocity in the tangential direction. E is often referred to as the Euler head and represents the ideal or theoretical head developed by the impeller only.

Now

$$C_{x1} = C_1 \cos \alpha_1 \quad \text{and} \quad C_{x2} = C_2 \cos \alpha_2$$

Thus

$$E = (U_2 C_2 \cos \alpha_2 - U_1 C_1 \cos \alpha_1)/g \tag{2.3}$$

But by using the cosine rule, $W^2 = U^2 + C^2 - 2UC \cos \alpha$, then

$$U_1 C_1 \cos \alpha_1 = (U_1^2 - W_1^2 + C_1^2)/2$$

and

$$U_2 C_2 \cos \alpha_2 = (U_2^2 - W_2^2 + C_2^2)/2$$

and substituting into Eq. (2.3) gives

$$E = [(U_2^2 - U_1^2) + (C_2^2 - C_1^2) + (W_1^2 - W_2^2)]/2g \qquad (2.4)$$

The terms in Eq. (2.4) may now be examined in turn.

$(C_2^2 - C_1^2)/2g$ represents the increase of kinetic energy of the fluid across the impeller, $(U_2^2 - U_1^2)/2g$ represents the energy used in setting the fluid into circular motion about the impeller axis and $(W_1^2 - W_2^2)/2g$ is the gain of static head due to a reduction of the relative velocity within the impeller. The flow rate is

$$Q = 2\pi r_1 C_{r1} b_1 = 2\pi r_2 C_{r2} b_2$$

where C_r is the radial component of the absolute velocity and is perpendicular to the tangent at inlet and outlet while b is the width of the blade (in the z direction). It is usually the case that $C_1 = C_{r1}$ and hence $\alpha_1 = 90°$. In this case $C_{x1} = 0$, where C_{x1} is the component of the inlet absolute velocity vector resolved into the tangential direction. W_x and C_x are often respectively called the relative and absolute whirl components of velocity. When $\beta_1 = \beta_1'$, this is referred to as the 'no-shock condition' at entry. In this case the fluid moves tangentially onto the blade. When $\beta_2 = \beta_2'$ there is no fluid slip at the exit.

2.3 SLIP FACTOR

It was stated in the previous section that the angle at which the fluid leaves the impeller, β_2', may not be the same as the actual blade angle β_2. This is due to fluid slip, and it occurs in both centrifugal pumps and centrifugal compressors, and manifests itself as a reduction in C_{x2} in the Euler pump equation. One explanation for slip is that of the relative eddy hypothesis.

Figure 2.5 shows the pressure distribution built up in the impeller passages due to the motion of the blades. On the leading side of a blade there is a high-pressure region while on the trailing side of the blade there is a low-pressure region, the pressure changing across the blade passage. This pressure distribution is similar to that about an aerofoil in a free stream and is likewise associated with the existence of a circulation around the blade, so that on the low-pressure side the fluid velocity is increased while on the high-pressure side it is decreased, and a non-uniform velocity distribution results at any radius. Indeed, the flow may separate from the suction surface of the blade. The mean direction of the flow leaving the impeller is therefore β_2' and not β_2 as is assumed in the zero-slip situation. Thus C_{x2} is reduced to C_{x2}' and the

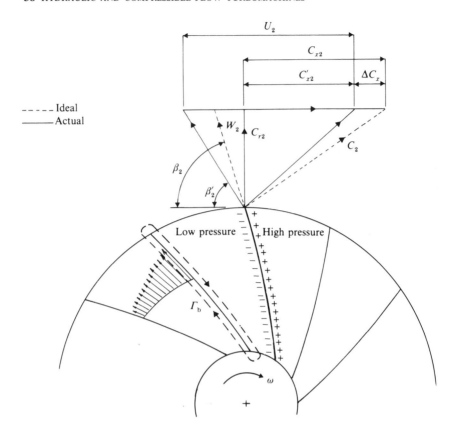

Figure 2.5 Slip and velocity distribution in centrifugal pump impeller blades

difference ΔC_x is defined as the slip. Slip factor is defined as

$$\text{Slip factor} = C'_{x2}/C_{x2} = \sigma_s \qquad (2.5)$$

Stodola[1] proposed the existence of a relative eddy within the blade passages as shown in Fig. 2.6. He proposed that if a frictionless fluid passes through the blade passages it will, by definition, have no rotation; therefore at the outlet of the passage rotation should be zero. Now the impeller has an angular velocity ω so that relative to the impeller the fluid must have an angular velocity $-\omega$ in the blade passages to comply with the zero-rotation condition. If the radius of a circle that may be inscribed between two successive blades at outlet and at a tangent to the surfaces of both blades is e, then the slip is given by

$$\Delta C_x = \omega e$$

Now the impeller circumference is $2\pi r_2$ and therefore the distance between

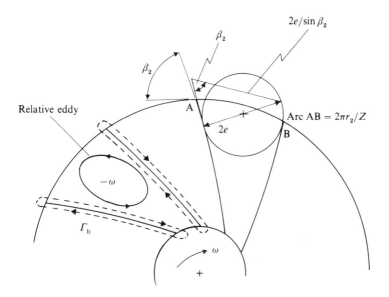

Figure 2.6 The relative eddy between impeller blades

blades is $2\pi r_2/Z$ if we have Z blades of negligible thickness. This may be approximated to $2e/\sin\beta_2$ and upon rearrangement

$$e = (\pi r_2/Z)\sin\beta_2$$

$$\Delta C_x = (U_2/Zr_2)(\pi r_2 \sin\beta_2)$$

$$= (U_2 \pi \sin\beta_2)/Z$$

Now referring back to Fig. 2.5 for the no-slip condition

$$C_{x2} = U_2 - C_{r2}\cot\beta_2$$

and substituting into Eq. (2.5) gives

$$\text{Slip factor} = (C_{x2} - \Delta C_x)/C_{x2}$$

$$= 1 - \Delta C_x/C_{x2}$$

$$= 1 - (U_2\pi \sin\beta_2)/[Z(U_2 - C_{r2}\cot\beta_2)]$$

$$= 1 - (\pi\sin\beta_2)/\{Z[1-(C_{r2}/U_2)\cot\beta_2]\} \quad (2.6)$$

For purely radial blades, which are often found in a centrifugal compressor, β_2 will be 90° and the Stodola slip factor becomes

$$\sigma_s = 1 - \pi/Z \quad (2.7)$$

The Stodola slip factor equation gives best results when applied in the range $20° < \beta_2 < 30°$. Other slip factors are named after Buseman[2]

$$\sigma_s = [A - B(C_{r2}/U_2)\cot \beta_2]/[1 - (C_{r2}/U_2)\cot \beta_2] \qquad (2.8)$$

where A and B are functions of β_2, Z and r_2/r_1, and are best used in the range $30° < \beta_2 < 80°$. The Stanitz[3] slip factor given by

$$\sigma_s = 1 - 0.63\pi/\{Z[1 - (C_{r2}/U_2)\cot \beta_2]\} \qquad (2.9)$$

is best used in the range $80° < \beta_2 < 90°$.

When applying a slip factor, the Euler pump equation (Eq. (1.25)) becomes

$$W/m = \sigma_s U_2 C_{x2} - U_1 C_{x1} \qquad (2.10)$$

Typically, slip factors lie in the region of 0.9, while slip occurs even if the fluid is ideal.

2.4 CENTRIFUGAL PUMP CHARACTERISTICS

In Sec. 1.3 dimensionless groups were used to express the power, head and flow relationships for a hydraulic machine. A well-designed pump should operate at or near the design point and hence near its maximum efficiency, but the engineer is often required to know how the pump will perform at off-design conditions. For instance, the head against which the pump is operating may be decreased, resulting in an increase in mass flow rate. However, before examining this aspect of off-design performance, we will look at the losses occurring in a pump and the differing efficiencies to which these losses give rise, whether or not the pump is working at the design point. We will then examine the effects of working at the off-design condition.

2.4.1 Pump Losses

The shaft power, P_s or energy that is supplied to the pump by the prime mover is not the same as the energy received by the liquid. Some energy is dissipated as the liquid passes through the machine and the mechanism of this loss can be split up into the following divisions.

1. Mechanical friction power loss, P_m due to friction between the fixed and rotating parts in the bearing and stuffing boxes.
2. Disc friction power loss, P_i due to friction between the rotating faces of the impeller (or disc) and the liquid.
3. Leakage and recirculation power loss, P_l due to a loss of liquid from the pump or recirculation of the liquid in the impeller.
4. Casing power loss, P_c

Impeller power loss is caused by an energy or head loss h_i in the impeller due to disc friction, flow separation and shock at impeller entry. This loss is

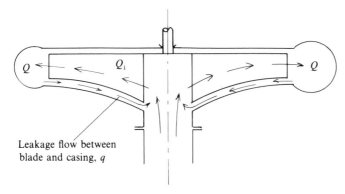

Leakage flow between
blade and casing, q

Figure 2.7 Leakage and recirculation in a centrifugal pump

associated with the flow rate Q_i through the impeller, and so the impeller power loss is expressed as

$$P_i = \rho g Q_i h_i \qquad (2.11)$$

However, while the flow through the impeller is Q_i, this is not the flow through the outlet or inlet flange of the machine. The pressure difference between impeller tip and eye can cause a recirculation of a small volume of fluid q, thus reducing the flow rate at outlet to Q as shown in Fig. 2.7, and then

$$Q = Q_i - q \qquad (2.12)$$

If H_i is the total head across the impeller, then a leakage power loss can be defined as

$$P_1 = \rho g H_i q \quad (\text{N m/s}) \qquad (2.13)$$

Equation (2.12) shows that when the discharge valve of the pump is closed, then the leakage flow rate attains its highest value.

In flowing from the impeller outlet to the pump outlet flange, a further head loss h_c takes place in the diffuser and collector, and since the flow rate here is Q, then a casing power loss may be defined as

$$P_c = \rho g Q h_c \quad (\text{N m/s}) \qquad (2.14)$$

Summing these losses gives

$$P_s = P_m + \rho g (h_i Q_i + h_c Q + H_i q + QH) \qquad (2.15)$$

where the total head delivered by the pump is defined as in Fig. 2.1 and Eq. (2.1).

A number of efficiencies are associated with these losses:

$$\text{Overall efficiency} = \frac{\text{Fluid power developed by pump}}{\text{Shaft power input}}$$

or

$$\eta_o = \rho g Q H / P_s \tag{2.16}$$

$$\text{Casing efficiency} = \frac{\text{Fluid power at casing outlet}}{\text{Fluid power at casing inlet}}$$

$$= \frac{\text{Fluid power at casing outlet}}{\text{Fluid power developed by impeller} - \text{Leakage loss}}$$

or

$$\eta_c = \rho g Q H / \rho g Q H_i = H / H_i \tag{2.17}$$

$$\text{Impeller efficiency} = \frac{\text{Fluid power at impeller exit}}{\text{Fluid power supplied to impeller}}$$

$$= \frac{\text{Fluid power at impeller exit}}{\text{Fluid power developed by impeller} + \text{Impeller loss}}$$

or

$$\eta_i = \rho g Q_i H_i / [\rho g Q_i (H_i + h_i)] = H_i / (H_i + h_i) \tag{2.18}$$

$$\text{Volumetric efficiency} = \frac{\text{Flow rate through pump}}{\text{Flow rate through impeller}}$$

or

$$\eta_v = Q / (Q + q) \tag{2.19}$$

$$\text{Mechanical efficiency} = \frac{\text{Fluid power supplied to the impeller}}{\text{Power input to the shaft}}$$

or

$$\eta_m = \rho g Q_i (h_i + H_i) / P_s \tag{2.20}$$

Therefore

$$\eta_o = \eta_c \eta_i \eta_v \eta_m \tag{2.21}$$

A hydraulic efficiency may be defined as

$$\eta_H = \frac{\text{Actual head developed by pump}}{\text{Theoretical head developed by impeller}}$$

$$= H / (H_i + h_i) \tag{2.22}$$

where the theoretical head $(H_i + h_i)$ is that obtained from Euler's equation (Eq. (1.25)) and $\eta_H = \eta_i \eta_c$.

Figure 2.8 shows how each of the power losses are subtracted from the initial input power. The rectangle OABC represents the total power input to the shaft while OADEFC is equivalent to the mechanical power loss. The impeller loss $\rho g Q_i h_i$ is next removed and is represented by rectangle EFGI. The

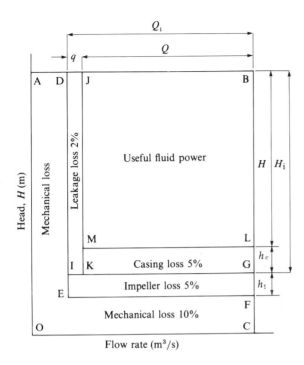

Figure 2.8 Losses in a centrifugal pump

next loss to be accounted for is the leakage loss $\rho g H_i q$ represented by rectangle DJKI, and finally the casing loss $\rho g h_c Q$ represented by rectangle MLGK is removed. This leaves us with rectangle JBLM, which represents the fluid power output or power developed by the pump $\rho g Q H$.

2.4.2 The Characteristic Curve

Euler's pump equation (Eq. (1.25)) gives the theoretical head developed by the pump, but if it is assumed that there is no whirl component of velocity at entry then $C_{x1} = 0$ and the actual theoretical head developed is

$$E = U_2 C_{x2}/g = (H_i + h_i) \qquad (2.23)$$

and if slip is accounted for, Eq. (2.23) becomes

$$E_N = \sigma_s E = \sigma_s (H_i + h_i) \qquad (2.24)$$

Now

$$C_{x2} = U_2 - W_{x2}$$

$$= U_2 - C_{r2} \cot \beta_2$$

$$= U_2 - (Q/A) \cot \beta_2$$

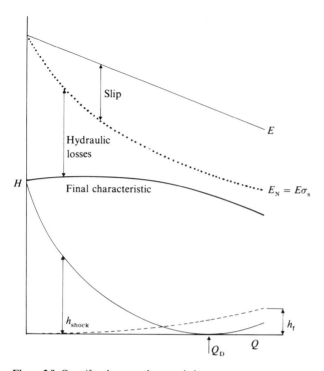

Figure 2.9 Centrifugal pump characteristic

where A is the flow area at the periphery of the impeller and C_r is perpendicular to it. Thus from Eq. (2.23) the energy per unit weight of flow becomes

$$E = U_2[U_2 - (Q/A)\cot\beta_2]/g$$

and since U_2, β_2 and A are constants, then

$$E = K_1 - K_2Q \qquad (2.25)$$

and this equation may be plotted as the straight line shown in Fig. 2.9.

If slip is taken into account, it is seen from Eq. (2.9) that as C_{r2} increases (and hence Q) then σ_s decreases, thus reducing the value of E in Eq. (2.25) to

$$E_N = (K_1 - K_2Q)\sigma_s \qquad (2.26)$$

The loss due to slip can occur in both a real and an ideal fluid, but in a real fluid account must also be taken of the shock losses at entry to the blades, and the friction losses in the casing and impeller vanes, or indeed at any point where the fluid is in contact with a solid surface of the pump. At the design point the shock losses are zero since the fluid would move tangentially onto the blade, but on either side of the design point the head loss due to shock increases according to

$$h_{\text{shock}} = K_3(Q - Q_D)^2 \qquad (2.27)$$

where Q_D is the design flow rate. The friction losses are accounted for in the form

$$h_f = K_4 Q^2 \qquad (2.28)$$

Equations (2.27) and (2.28) are plotted in Fig. 2.9 and the sum of them is subtracted from the curve of Eq. (2.26) to give the final characteristic. This curve is called the head–flow characteristic of the pump.

2.4.3 Effect of Flow Rate Variation

A pump is usually designed to run at a fixed speed with a design head and flow rate and these conditions would normally occur at the maximum efficiency point. However, it is not always the case in practice that the operating point lies at the design point. This may be due to a pipeline being partially blocked, a valve jammed partially closed or poor matching of the pump to the piping system. Also in general a variable-speed motor is not available to correct for any deviation from the design condition, so that in what follows it is assumed that the speed of the pump remains constant.

Figure 2.10 shows the velocity diagrams that pertain for three possible flow rates: normal design flow rate, increased flow rate and decreased flow rate. When the flow rate changes, C_{r2} changes, and since U_2 is constant and the blade outlet angle β_2 is constant (assuming $\beta_2' = \beta_2$), the magnitude of W_2 and C_2 must change along with the angle α_2. Since the effective energy transfer E depends on C_{x2}, then E will change accordingly. Thus a reduction in Q gives an increase in C_{x2}, while an increase in Q gives a reduction in C_{x2}. It follows that, should the head against which the pump operates be momentarily increased, E and therefore C_{x2} increase and Q decreases to give the new operating point at the increased head. Similarly a reduction in the operating head gives an increase in Q.

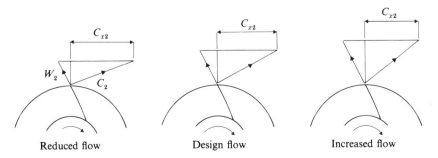

| Reduced flow | Design flow | Increased flow |

Figure 2.10 Effect of flow rate variation on outlet velocity

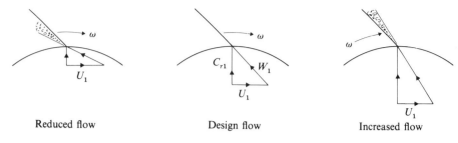

| Reduced flow | Design flow | Increased flow |

Figure 2.11 Effect of flow rate variation on inlet velocity

At the inlet the effect of flow rate change is to cause eddies on the suction surface of the blade for a reduced flow rate and on the pressure surface of the blade for an increased flow rate. The design condition is the 'no-shock' condition, which corresponds to the flow Q_D in Fig. 2.9. The corresponding velocity diagrams can be seen in Fig. 2.11. In all cases it is assumed that C_{x1} is zero.

2.4.4 Effect of Blade Outlet Angle

The characteristic curve will also be affected by the blade angle at outlet, the three types of blade settings being backward-facing, forward-facing and radial blades. Figure 2.12 shows clearly the velocity triangles for each case with $C_{x1} = 0$.

Case (i). Backward-facing blades, $\beta_2 < 90°$

$$C_{x2} = U_2 - C_{r2} \cot \beta_2$$

Therefore

$$E = U_2(U_2 - C_{r2} \cot \beta_2)/g$$

or

$$E = (U_2^2/g) - (Q U_2 \cot \beta_2/gA)$$

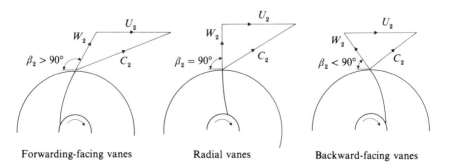

| Forwarding-facing vanes | Radial vanes | Backward-facing vanes |

Figure 2.12 Centrifugal pump outlet velocity triangles for varying blade outlet angle

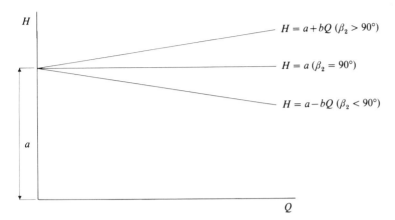

Figure 2.13 Theoretical characteristics for varying outlet blade angle

Writing E as a head,

$$H = a - bQ \qquad (2.29)$$

and for pumps β_2 typically lies between $15°$ and $90°$.

Case (ii). Radial blades, $\beta_2 = 90°$

$$H = a \qquad (2.30)$$

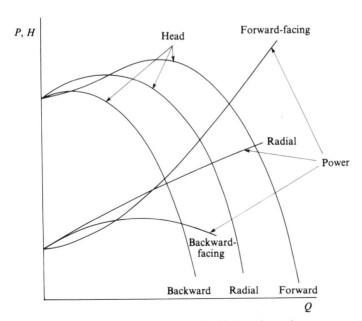

Figure 2.14 Actual characteristics for varying blade outlet angle

Case (iii). Forward-facing vanes, $\beta_2 > 90°$

$$H = a + bQ \qquad (2.31)$$

where β_2 would be typically $140°$ for a multi-bladed centrifugal fan.

These equations are plotted in Fig. 2.13 as characteristics and they revert to their more recognized curved shapes (for the reasons previously discussed) as shown in Fig. 2.14.

For both radial and forward-facing blades the power is rising continuously as the flow rate is increased. In the case of backward-facing vanes the maximum efficiency occurs in the region of maximum power, and if, for some reason, Q increases beyond Q_D, this results in a power decrease and therefore the motor used to drive the pump may be safely rated at the maximum power. This is said to be a self-limiting characteristic. In the case of the radial and forward-facing vanes, if the pump motor is rated for maximum power, then it will be under-utilized most of the time, and extra cost will have been incurred for the extra rating, whereas if a smaller motor is employed rated at the design point, then if Q increases above Q_D the motor will be overloaded and may fail. It therefore becomes more difficult to decide on a choice of motor for these latter cases.

2.5 FLOW IN THE DISCHARGE CASING

The discharge casing is that part of the casing following the impeller outlet. It has two functions: (i) to receive and guide the liquid discharged from the impeller to the outlet ports of the pump, and (ii) to increase the static head at the outlet of the pump by reducing the kinetic energy of the liquid leaving the impeller. These two functions may be called collector and diffuser functions. The former function may be used alone while the latter can occur either before or after the collector function. In addition diffusion can take place in a vaned or vaneless diffuser.

2.5.1 Volute or Scroll Collector

A simple volute or scroll collector is illustrated in Fig. 2.15 and consists of a circular passage of increasing cross-sectional area. The advantage of the simple volute is its low cost. The cross-sectional area increases as the increment of discharge increases around the periphery of the impeller and it is found that a constant average velocity around the volute results in equal pressures around the pump casing, and hence no radial thrust on the shaft. Any deviation in capacity (flow rate) from the design condition will result in a radial thrust, which if allowed to persist could result in shaft bending. Values of radial thrust are given by the empirical relationship[4]

$$P = 495\,KHD_2B_2 \qquad (2.32)$$

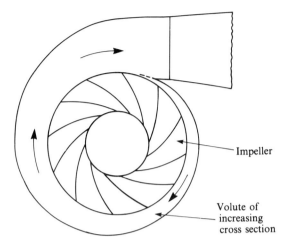

Impeller

Volute of
increasing
cross section

Figure 2.15 Simple volute or scroll collector

where P = radial force (N), H = head (m), D_2 = peripheral diameter (m), B_2 = impeller width (m) and K = constant determined from the following equation for a particular value of Q:

$$K = 0.36[1 - (Q/Q_D)^2] \qquad (2.33)$$

A cross section of the volute casing is shown in Fig. 2.16. The circular section is adopted to reduce the losses due to friction and impact when the fluid hits the casing walls on exiting from the impeller. Of the available kinetic energy at impeller outlet, 25–30 per cent may be recovered in a simple volute.

2.5.2 Vaneless Diffuser

Diffusion takes place in a parallel-sided passage and is governed by the principle of conservation of angular momentum of the fluid. The outlet

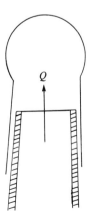

Q

Figure 2.16 Section through volute casing

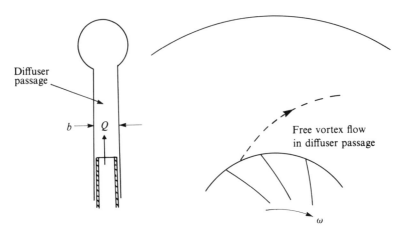

Figure 2.17 Vaneless diffuser passage

tangential velocity is reduced as the radius increases, while the radial component of absolute velocity is controlled by the radial cross-sectional area of flow b. A vaneless diffuser passage is shown in Fig. 2.17.

With reference to Fig. 2.18 the size of the diffuser may be determined as follows. The mass flow rate m at any radius r is given by

$$m = \rho A C_r = 2\pi r b \rho C_r$$

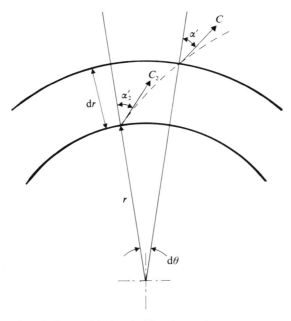

Figure 2.18 Logarithmic spiral flow in vaneless space

where b is the width of the diffuser passage perpendicular to the peripheral area of the impeller and is usually the same as the impeller width. Letting the subscripted variables represent conditions at the impeller outlet and the unsubscripted variables represent conditions at any radius r in the vaneless diffuser, then from continuity

$$rb\rho C_r = r_2 b_2 \rho_2 C_{r2}$$

or

$$C_r = r_2 b_2 \rho_2 C_{r2}/rb\rho \qquad (2.34)$$

If frictionless flow is assumed, then by the conservation of angular momentum

$$C_x = C_{x2} r_2/r$$

But $C_x \gg C_r$ (usually) and therefore the absolute velocity C is approximately equal to C_x or

$$C = C_{x2} r_2/r \qquad (2.35)$$

From Eq. (2.35), for C to be small, which is what we are trying to achieve, then r must be large and therefore, for a large reduction in the outlet kinetic energy, a diffuser with a large radius is required.

For an incompressible fluid, the inclination of the absolute velocity vector to the radial line remains constant at all θ since at the outlet from the impeller (Fig. 2.18)

$$\tan \alpha_2' = C_{x2}/C_{r2} = \text{constant} = \tan \alpha'$$

since rC_r is constant from the constant mass flow rate requirement, and $C_x r$ is constant from the conservation of angular momentum requirement. Thus the flow in the diffuser remains at a constant inclination α' to radial lines, the flow path tracing out a logarithmic spiral, and if for an incremental radius dr the fluid moves through angle $d\theta$, then from Fig. 2.18

$$r \, d\theta = (\tan \alpha') \, dr$$

Integrating,

$$\theta - \theta_2 = (\tan \alpha') \log_e(r/r_2) \qquad (2.36)$$

Putting $\alpha' = 78°$ and $(r/r_2) = 2$, the change in angle of the diffuser is almost $180°$, giving rise to a long flow path, which may result in high frictional losses, which in turn gives a low efficiency. So it is seen that the length of the diffuser must be balanced by the pressure recovery that is required and an optimum point is usually found based on either economic or hydraulic friction loss considerations.

2.5.3 Vaned Diffuser

The vaned diffuser shown in Fig. 2.19 is able to diffuse the outlet kinetic energy at a much higher rate, in a shorter length and with a higher efficiency than the

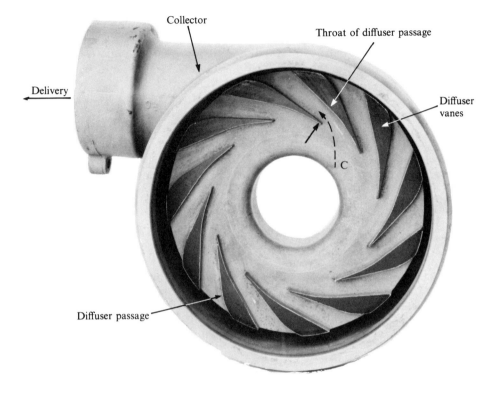

Figure 2.19 A vaned diffuser

vaneless diffuser. This is very advantageous where the size of the pump is important.

A ring of diffuser vanes surrounds the impeller at the outlet, and after leaving the impeller the fluid moves in a logarithmic spiral across a short vaneless space before entering the diffuser vanes proper. Once the fluid has entered the diffuser passage, the controlling variable on the rate of diffusion is the divergence angle of the diffuser passage, which is of the order of 8–10° and should ensure no boundary-layer separation along the passage walls. The number of vanes on the diffuser ring is subject to the following considerations:

1. The greater the vane number, the better is the diffusion but the greater is the friction loss.
2. The cross section of the diffuser channel should be square to give a maximum hydraulic radius (cross-sectional area/channel perimeter).
3. The number of diffuser vanes should have no common factor with the number of impeller vanes. This is to obviate resonant or sympathetic vibration.

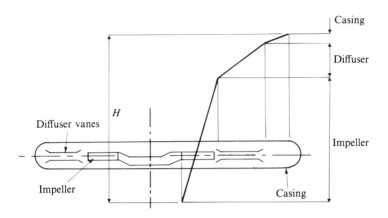

Figure 2.20 Head rise across a centrifugal pump

The collector and diffuser operate at their maximum efficiency at the design point only. Any deviation from the design discharge will alter the outlet velocity triangle and the subsequent flow in the casing. Figure 2.20 shows the contribution of each section of the pump to the total head developed by the pump.

2.6 CAVITATION IN PUMPS

Cavitation is a phenomenon that occurs when the local absolute static pressure of a liquid falls below the vapour pressure of the liquid and thereby causes vapour bubbles to form in the main body of liquid, that is the liquid boils. When the liquid flows through a centrifugal or axial flow pump, the static pressure (suction pressure) at the eye of the impeller is reduced and the velocity increases. There is therefore a danger that cavitation bubbles may form at the inlet to the impeller. When the fluid moves into a higher-pressure region, these bubbles collapse with tremendous force, giving rise to pressures as high as 3500 atm. Local pitting of the impeller can result when the bubbles collapse on a metallic surface, and serious damage can occur from this prolonged cavitation erosion, as shown in Fig. 2.21. Noise is also generated in the form of sharp cracking sounds when cavitation takes place.

Referring to Fig. 2.1, cavitation is most likely to occur on the suction side of the pump between the lower reservoir surface and the pump inlet since it is in this region that the lowest pressure will occur. A cavitation parameter σ is defined as

$$\sigma = \frac{\text{Pump total inlet head above vapour pressure}}{\text{Head developed by pump}}$$

Figure 2.21 Cavitation erosion in centrifugal pump impeller

and at the inlet flange

$$\sigma = (p_i/\rho g + V_i^2/2g - p_{vap}/\rho g)/H \qquad (2.37)$$

where all pressures are absolute. The numerator of Eq. (2.37) is a suction head and is called the net positive suction head ($NPSH$) of the pump. It is a measure of the energy available on the suction side of the pump.

Every pump has a critical cavitation number σ_c, which can only be determined by testing to find the minimum value of $NPSH$ before cavitation occurs. Various methods exist for determining the point of cavitation inception, and σ_c, and therefore the minimum $NPSH$ required by the pump, will depend upon the criteria chosen to define σ_c as well as the conditions under which the test is carried out. One method is to determine the normal head–flow characteristic of the pump and then to repeat the test with the inlet to the pump progressively throttled so as to increase the resistance to flow at the inlet. It will be found that for different throttle valve settings the performance curve will fall away from the normal operating curve at various points and one definition of the occurrence of minimum $NPSH$ is the point at which the head H drops below the normal operating characteristic by some arbitrarily selected percentage, usually about 3 per cent. At this condition, static inlet pressure p_i and inlet velocity V_i are measured, and σ_c is then calculated from Eq. (2.37). The minimum required $NPSH$ or σ_c may then be plotted for the different degrees of inlet throttling to give a curve of σ_c versus flow coefficient (Fig. 2.22).

In Fig. 2.1, the energy loss between the free surface (A) and the inlet side of the pump (i) is given by the steady flow energy equation as

Energy at A − Energy at i = Energy lost between A and i

Writing the energy in terms of heads

$$(p_A/\rho g + V_A^2/2g + Z_A) = (p_i/\rho g + V_i^2/2g + Z_i) + (h_{in} + h_{fi}) \qquad (2.38)$$

where $(h_{in} + h_{fi})$ represents the losses. Now V_A equals zero and, if the datum is

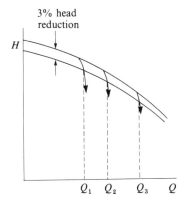

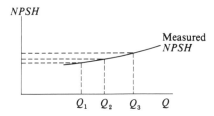

Figure 2.22 Critical $NPSH$ plotted on the pump characteristic

placed at the lower reservoir surface, then Z_A is also zero and Eq. (2.38) becomes

$$p_A/\rho g = p_i/\rho g + V_i^2/2g + H_{\text{suction}}$$

where

$$Z_i + h_{\text{in}} + h_{\text{fi}} = H_{\text{suction}}$$

Substituting for $p_i/\rho g$ in Eq. (2.37) gives

$$\sigma = (p_A/\rho g - p_{\text{vap}}/\rho g - H_{\text{suction}})/H \qquad (2.39)$$

Providing σ is above σ_c, cavitation will not occur, but, in order to achieve this, it may be necessary to decrease H_{suction} by decreasing Z_i and in some cases the pump may have to be placed below the reservoir or pump free surface, i.e. negative Z_i, especially if h_{fi} is particularly high due to a long inlet pipe. Thus when the pump is connected to any other inlet pipe system, σ as determined from Eq. (2.39) may be calculated and providing σ(available) $> \sigma_c$(required) then cavitation will be avoided.

2.6.1 Suction Specific Speed

It is reasonable to expect that the efficiency will be dependent not only upon the flow coefficient but also upon another function due to cavitation. The other function is the dimensionless suction specific speed and is defined as

$$N_{suc} = NQ^{1/2}/[g(NPSH)]^{3/4} \qquad (2.40)$$

Thus

$$\eta = f(\phi, N_{suc}) \qquad (2.41)$$

It is found from experiments[2] that the inception of cavitation occurs at constant values of N_{suc} and empirical results show that $N_{suc} \approx 3$ for N in rad/s, Q in m³/s and $g(NPSH)$ in m²/s². The cavitation parameter may also be determined by dividing the dimensionless specific speed by the dimensionless suction specific speed:

$$\begin{aligned} N_s/N_{suc} &= [NQ^{1/2}/(gH)^{3/4}]/\{NQ^{1/2}/[g(NPSH)]^{3/4}\} \\ &= (NPSH)^{3/4}/H^{3/4} \\ &= \sigma_c^{3/4} \end{aligned} \qquad (2.42)$$

Also from the similarity laws

$$NPSH_1/NPSH_2 = (N_1/N_2)^2(D_1/D_2)^2 = \sigma_1/\sigma_2$$

2.7 AXIAL FLOW PUMP

An axial flow pump consists of a propeller type of impeller running in a casing with fine clearances between the blade tips and the casing walls. In the absence of secondary flows, fluid particles do not change radius as they move through the pump; however, a considerable amount of swirl in the tangential direction will result unless means are provided to eliminate the swirl on the outlet side. This is usually done by fitting outlet guide vanes. The flow area is the same at

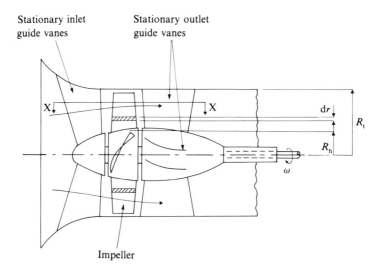

Figure 2.23 An axial flow pump

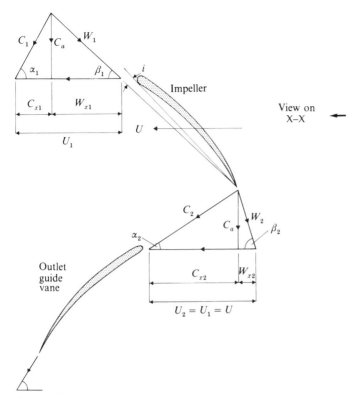

Figure 2.24 Axial flow pump velocity triangles

inlet and outlet and the maximum head for this type of pump is of the order of 20 m. It may be seen in Fig. 1.7 that the dimensionless specific speed of axial flow pumps lies at the right-hand side of the pump spectrum, its characteristics being one of low head but high capacity. The usual number of blades lies between two and eight, with a hub diameter/impeller diameter ratio of 0.3–0.6. In many cases the blade pitch is fixed but most large hydroelectric units have variable-pitch blades to allow for load variations.

Figure 2.23 shows an axial flow pump impeller. The section through the blade at X–X is shown enlarged with the inlet and outlet velocity triangles superimposed in Fig. 2.24. It will be noticed that the blade has an aerofoil section and that the inlet relative velocity vector W_1 does not impinge tangentially but rather the blade is inclined at an angle of incidence i to the relative velocity vector W_1. This is similar to the angle of attack of an aerofoil in a free stream. It is assumed that there is no shock at entry and that the fluid leaves the blade tangentially at exit.

Changes in the condition of the fluid take place at a constant mean radius; therefore

$$U_1 = U_2 = U = \omega r$$

Assuming also a constant flow area from inlet to outlet

$$C_{r1} = C_{r2} = C_r = C_a$$

and noting that the flow area is the annulus formed between the hub and the blade tips, then we may write

$$m = \rho C_a \pi (R_t^2 - R_h^2)$$

From Eq. (2.2),

$$E = U(C_{x2} - C_{x1})/g \qquad (2.43)$$

and for maximum energy transfer $C_{x1} = 0$, i.e. $\alpha_1 = 90°$ and $C_1 = C_a$, the absolute flow velocity at inlet being axial for maximum energy transfer.

Now

$$\cot \beta_2 = (U - C_{x2})/C_a$$

or

$$C_{x2} = U - C_a \cot \beta_2$$

Hence substituting for C_{x2} in Eq. (2.43) with $C_{x1} = 0$, the maximum energy transfer or head is

$$E = U(U - C_a \cot \beta_2)/g \qquad (2.44)$$

For constant energy transfer, Eq. (2.44) applies over the whole span of the blade from hub to tip; that is it applies at any radius r between R_t and R_h. For E to be constant over the whole blade length it is obvious that, as U^2 increases with radius, so an equal increase in $UC_a \cot \beta_2$ must take place and since C_a is constant then $\cot \beta_2$ must increase, and the blade must therefore be twisted as the radius changes.

Strictly speaking the work done per unit weight of flow through an annulus of thickness dr should be considered and this then integrated across the whole flow area from the hub to the tip

$$E = W/mg = U(U - C_a \cot \beta_2)/g$$

or

$$dW = U(U - C_a \cot \beta_2)dm$$

But the incremental mass flow rate dm is

$$dm = \rho(2\pi r)(C_a)\,dr$$

and $U = \omega r$. Therefore

$$W = 2\pi \rho C_a \omega \int_{R_h}^{R_t} r^2(\omega r - C_a \cot \beta_2)\,dr \qquad (2.45)$$

Equation (2.45) can only be integrated if the relationship between β_2 and r is known.

For design purposes it is usual to select conditions for use in Eq. (2.44) at the mean radius $(R_h + R_t)/2$ along the blade. The whirl component imparted to the fluid at exit from the blade is relatively small, resulting in a low kinetic energy loss. An axial flow pump therefore tends to have a higher hydraulic efficiency than the centrifugal pump.

2.7.1 Blade Element Theory

An axial flow pump impeller may have a large number of blades spaced closely together or a few blades spaced far apart, while for mechanical strength considerations, the blade chord will vary from hub to tip. The peripheral distance between similar points on two adjacent blades is the pitch, and the ratio blade chord/blade pitch at a given radius is known as the solidity ratio σ:

$$\sigma = c/s \qquad (2.46)$$

It is therefore possible to have high- or low-solidity blades, an impeller with a low number of blades implying a low solidity. Where the blades have a low solidity, flow interference from one blade to the next is low and the blade may be considered to be acting alone in a free stream and is analysed as such. However, for high-solidity blades implying very closely spaced blades, the flow between the blades will be greatly influenced by the adjacent blades and we must resort to cascade data for an analysis of the forces acting on them. Since axial flow pump impellers invariably have less than six blades, it is usual to consider only isolated blade element theory for them and this is now briefly

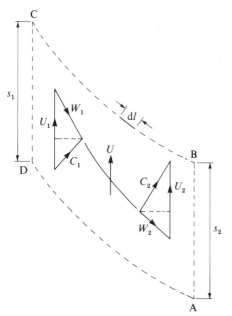

Figure 2.25 Circulation around an isolated blade

described, the treatment of cascade analysis being reserved for the section on axial flow compressors, to which machines it is more appropriately applied.

Consider the circulation Γ around the control surface of the isolated blade shown in Fig. 2.25, where the lengths AB and CD are the blade pitches at inlet and outlet and AD and BC bisect the flow passages between adjacent blades.

The circulation is given by the line integral around ABCD and may be evaluated by summing the individual circulations comprising the circuit, such that

$$\Gamma_{ABCD} = \oint V \, ds = \int_A^B V \, ds_2 + \int_B^C V \, dl + \int_C^D V \, ds_1 + \int_D^A V \, dl \quad (2.47)$$

where the circulation is positive anticlockwise. But

$$\int_B^C V \, dl = -\int_D^A V \, dl$$

while

$$\int_A^B V \, ds_2 = C_{x2}s_2 \quad \text{and} \quad \int_C^D V \, ds_1 = -C_{x1}s_1$$

Hence

$$\Gamma_{ABCD} = s_2 C_{x2} - s_1 C_{x1} \quad (2.48)$$

For a number of blades shown in Fig. 2.26 the circulation around each blade may be determined, and the total circulation about Z blades is the sum of the individual circulations, remembering that along a line such as BG the circulation for one blade is positive (anticlockwise) while for the adjacent blade it is negative (clockwise). Therefore for Z blades the total circulation is

$$\Gamma_{ADEH} = Z\Gamma_b \quad (2.49)$$

where Γ_b is the circulation around a single blade.

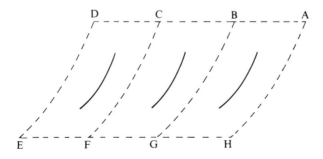

Figure 2.26 Circulation around a number of isolated blades

Substitution into Eq. (2.48) gives the circulation for the whole impeller as

$$\Gamma = Z(s_2 C_{x2} - s_1 C_{x1})$$

But

$$Zs_1 = 2\pi r_1 \quad \text{and} \quad Zs_2 = 2\pi r_2$$

Therefore

$$\Gamma = 2\pi(r_2 C_{x2} - r_1 C_{x1})$$

From Eq. (2.43),

$$E = \omega(r_2 C_{x2} - r_1 C_{x1})/g$$

Hence

$$E = \omega Z\Gamma_b/2\pi g \qquad (2.50)$$

The Kutta Joukowski law ($L = \rho U_0 \Gamma_b$) for lift per unit span on an aerofoil may now be used, where U_0 is the free stream velocity. Dividing this by $0.5\rho U_0^2 c$ gives the lift coefficient

$$C_L = \Gamma_b/0.5 U_0 c$$

where c is the chord of the impeller blade.

Since the appropriate free stream velocity for flow over the blade is the relative velocity W instead of U_0, and since this is different at blade inlet and outlet, the appropriate relative velocity is usually defined as

$$W_\infty^2 = C_a^2 + [(W_{x1} + W_{x2})/2]^2$$
$$= C_a^2 + [\cot \beta_1 + \cot \beta_2)/2]^2$$

Therefore $\Gamma_b = 0.5 C_L W_\infty c$, and substituting this into Eq. (2.50)

$$E = 0.5\omega Z C_L W_\infty c/2\pi g \qquad (2.51)$$

Values of lift coefficients for differing blade profiles may be determined from readily available tables and charts[5] and an estimate for E obtained.

2.7.2 Axial Flow Pump Characteristics

Axial flow pump design has evolved empirically and it is only in relatively modern times that aerofoil theory has been applied. Nevertheless, efficiencies of over 90 per cent were achieved using empirical data and it would seem that aerodynamic design has not improved the efficiencies by much.

Typical head–flow, power and efficiency curves are shown in Fig. 2.27. A steep negative slope is evident on the head and power curves at low flow rates. This can be explained by considering Eq. (2.44). For a given blade design at fixed speed with axial flow at inlet

$$E = U(U - C_a \cot \beta_2)/g$$

Now Q is proportional to C_a and therefore

$$dE/dC_a \propto dE/dQ \propto - U \cot \beta_2$$

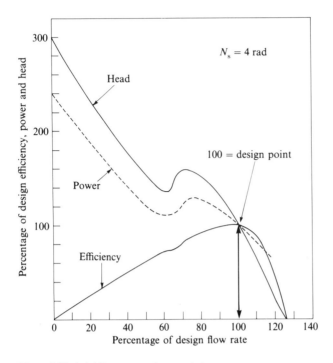

Figure 2.27 Axial flow pump characteristics

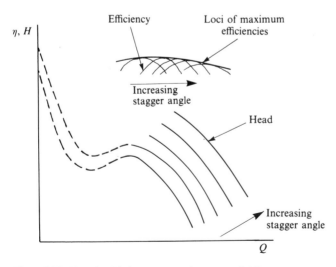

Figure 2.28 Changing blade stagger angle on an axial flow pump

62

For axial flow at inlet, β_2 is relatively small, and thus for a given pump at a given speed the head–flow relationship has a steep negative slope. The power curve is similarly very steep, the power requirement at shut-off being perhaps 2–2.5 times that required at the design point. This makes for a very expensive electric motor to cover the eventuality of low flow rates and so the fixed-blade axial flow pump is limited to operation at the fixed design point. Variable-flow machines may be designed employing variable blade stagger or setting angles. Here the blade angle is adjusted so that the pump runs at its maximum efficiency at all loads and also reduces the shut-off power requirement. Figure 2.28 shows the effect of changing blade stagger angle.

In Fig. 2.27 the power and head curves are seen to enter a region of instability at about 50 per cent of the design flow rate. This is due to C_a becoming increasingly small and thereby increasing the angle of incidence of flow onto the blade until separation and stalling of the blade occurs. The further head rise at even lower flow rates and the consequent power increase is due to recirculation of the fluid around the blade from the pressure side to the suction side and then up onto the pressure side of the next blade. An increased blade stagger angle will once again reduce this recirculation and thereby the power requirement.

2.8 PUMP AND SYSTEM MATCHING

It has been shown that a hydraulic pump has a design point at which the overall efficiency of operation is a maximum. However, it may happen that the pipe system in which the pump is being used is unsuited to the pump and a different pump with a more suitable characteristic is required. This section will examine how a pump and a pipe system may be matched to each other, the effect of changing the pump speed and diameter, and finally the effect of connecting pumps in series and parallel.

Consider the pipe system in Fig. 2.1. On the suction side the losses expressed in terms of standard loss coefficients are the sum of the minor losses h_{in}

$$h_{in} = \sum k V_i^2 / 2g$$

and the friction loss

$$h_{fi} = 4 f l_i V_i^2 / 2 g d_i$$

where f is the Darcy friction factor, l_i is the length of the inlet pipe and d_i its diameter. Thus the total head loss is

$$h_i = 4 f l_i V_i^2 / 2 g d_i + \sum k V_i^2 / 2g$$

On the delivery side the sum of the bend, friction and exit losses that must be overcome is

$$h_o = 4fl_oV_o^2/2gd_o + \sum kV_o^2/2g$$

Finally, the liquid must be moved from the lower reservoir to the upper reservoir through the static head H_s; hence the total opposing head of the pipe system that must be overcome in order to move the fluid from the lower to upper reservoir is

$$H = H_s + h_o + h_i \qquad (2.52)$$

Now from the continuity equation (Eq. (1.21)) the flow rate through the system is proportional to the velocity. Thus the resistance to flow in the form of friction losses, head losses, etc., is proportional to the square of the flow rate and is usually written as

$$\text{System resistance} = KQ^2 \qquad (2.53)$$

It is a measure of the head lost for any particular flow rate through the system. If any parameter in the system is changed, such as adjusting a valve opening, or inserting a new bend, etc., then K will change. The total system head loss of Eq. (2.52) therefore becomes

$$H = H_s + KQ^2 \qquad (2.54)$$

and if this equation is plotted on the head–flow characteristic, the point at which Eq. (2.54) intersects the pump characteristic is the operating point, and this may or may not lie at the duty point, which usually corresponds to the design point and maximum efficiency. The closeness of the operating and duty points depends on how good an estimate of the expected system losses has been made. In Fig. 2.29 the system curve is superimposed on the H–Q characteristic.

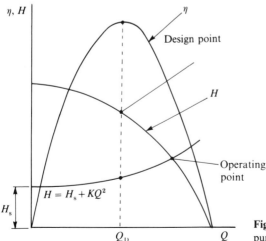

Figure 2.29 System resistance and pump characteristic curves

It should be noted that if there is no static head rise of the liquid (e.g. pumping in a horizontal pipeline between two reservoirs at the same elevation) then H_s is zero and the system curve passes through the origin. This has implications when speed and diameter changes take place. Because of the flatness of rotodynamic pump characteristics, a poor estimate of the system losses can seriously affect the flow rate and head; whereas in positive displacement pumps, the H–Q curve is almost vertical and, even if the head changes substantially, the flow rate stays almost constant.

2.8.1 Effect of Speed Variation

Consider a pump of fixed diameter pumping liquid with zero static lift. If the characteristic at one speed N_1 is known, then it is possible to predict the corresponding characteristic at speed N_2 and also the corresponding operating points. Figure 2.30 shows the characteristic at speed N_1.

For points A, B and C the corresponding head and flows at a new speed N_2 are found thus. We have

$$\phi_1 = \phi_2$$

and

$$Q_1/N_1 = Q_2/N_2 \quad \text{since } D \text{ is constant} \quad (2.55)$$

Similarly

$$\psi_1 = \psi_2$$

and

$$H_1/N_1^2 = H_2/N_2^2 \quad (2.56)$$

Applying Eqs (2.55) and (2.56) to points A, B and C and letting the corresponding points be A′, B′ and C′,

$$Q_2 = Q_1 N_2/N_1 \quad \text{and} \quad H_2 = H_1(N_2)^2/(N_1)^2$$

or

$$H_2 \propto Q_2^2 \quad (2.57)$$

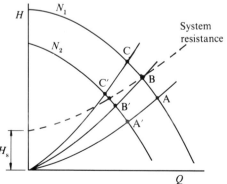

Figure 2.30 Effect of speed variation on the operating point

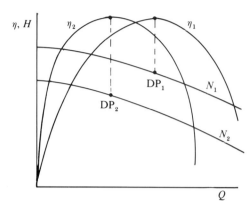

Figure 2.31 Effect of speed variation on the efficiency

and we see that all corresponding points lie on a parabola passing through the origin. This means that for an operating point at A at speed N_1, it is only necessary to apply the similarity laws directly to find the corresponding operating point at the new speed since it will lie on the system curve itself. However, if there is static lift ($H_s \neq 0$) it will be necessary to calculate and then plot the corresponding points A', B', C' at the new speed, since the system curve will no longer pass through the origin. The system curve is then drawn to find the new operating point at its intersection with the N_2 characteristic such that the corresponding maximum efficiency at design point DP_2 remains the same as at DP_1 but at a different head and flow rate as in Fig. 2.31.

2.8.2 Variation of Pump Diameter

A variation of pump diameter may be similarly examined through the similarity laws. For a constant speed,

$$Q_1/D_1^3 = Q_2/D_2^3$$

and

$$H_1/D_1^2 = H_2/D_2^2$$

or

$$H \propto Q^{2/3} \tag{2.58}$$

This curve does not lie on the system characteristic and therefore part of the new characteristic must be drawn through A', B' and C' at the new diameter so that the new operating point may be found. The efficiency curve moves across in a similar manner to before, the corresponding efficiencies being equal.

2.8.3 Pumps in Series and Parallel

Should the head or flow rate of a single pump not be sufficient for an application, pumps can be combined in series to obtain an increase in head, or

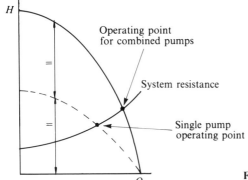

Figure 2.32 Two similar pumps in series

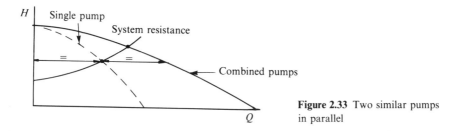

Figure 2.33 Two similar pumps in parallel

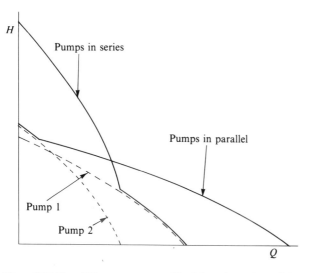

Figure 2.34 Two different pumps combined in series and parallel

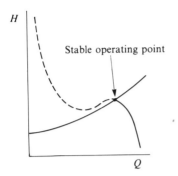

Figure 2.35 Single axial flow pump

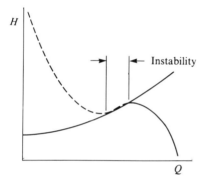

Figure 2.36 Axial flow pumps in parallel

in parallel for an increase in flow rate. The combined pumps need not be of the same design.

Figures 2.32 and 2.33 show the combined H–Q characteristic for the cases of identical pumps connnected in series and parallel. It will be observed that the operating point changes in both cases. In Fig. 2.34 the combined characteristics of two different pumps connected in parallel and series are drawn.

When axial flow pumps are connected in parallel, care must be taken to ensure that the system characteristic does not cut the pump characteristic in two places, otherwise instability may result. This arises due to the lateral spreading of the pump characteristic illustrated in Figs 2.35 and 2.36.

EXERCISES

2.1 A centrifugal pump of 1.3 m diameter delivers 3.5 m³/min of water at a tip speed of 10 m/s and a flow velocity of 1.6 m/s. The outlet blade angle is at 30° to the tangent at the impeller periphery. Assuming zero whirl at inlet, and zero slip, calculate the torque delivered by the impeller.

2.2 The impeller of a centrifugal pump has backward-facing blades inclined at 30° to the tangent at impeller outlet. The blades are 20 mm in depth at the outlet, the impeller is 250 mm in diameter and it rotates at 1450 rpm. The flow rate through the pump is 0.028 m^3/s and a slip factor of 0.77 may be assumed. Determine the theoretical head developed by the impeller, and the number of impeller blades.

2.3 An impeller with an eye radius of 51 mm and an outside diameter of 406 mm rotates at 900 rpm. The inlet and outlet blade angles measured from the radial flow direction are 75° and 83° respectively, while the blade depth is 64 mm. Assuming zero inlet whirl, zero slip and an hydraulic efficiency of 89 per cent, calculate

    (a) the volume flow rate through the impeller,
    (b) the stagnation and static pressure rise across the impeller,
    (c) the power transferred to the fluid and
    (d) the input power to the impeller.

2.4 The basic design of a centrifugal pump has a dimensionless specific speed of 0.075 rev. The blades are forward facing on the impeller and the outlet angle is 120° to the tangent, with an impeller passage width at outlet equal to one-tenth of the diameter. The pump is to be used to pump water a vertical distance of 35 m at a flow rate of 0.04 m^3/s. The suction and delivery pipes are each of 150 mm diameter and have a combined length of 40 m with a friction factor of 0.005. Other losses at pipe entry, exit, bends, etc., are three times the velocity head in the pipes. If the blades occupy 6 per cent of the circumferential area and the hydraulic efficiency (neglecting slip) is 76 per cent, what must be the diameter of the pump impeller?

2.5 When a laboratory test was carried out on a pump, it was found that, for a pump total head of 36 m at a discharge of 0.05 m^3/s, cavitation began when the sum of the static pressure plus the velocity head at inlet was reduced to 3.5 m. The atmospheric pressure was 750 mmHg and the vapour pressure of water 1.8 kPa. If the pump is to operate at a location where atmospheric pressure is reduced to 620 mmHg and the vapour pressure of water is 830 Pa, what is the value of the cavitation parameter when the pump develops the same total head and discharge? Is it necessary to reduce the height of the pump above the supply, and if so by how much?

2.6 The inner and outer diameters of an axial flow pump are 0.75 and 1.8 m respectively. Fixed stator blades lie downstream of the rotor with an inlet angle of 40° (at the mean diameter) measured from the direction of blade motion. The rotor blade outlet angle (at the mean diameter) also measured from the direction of blade motion is 30° and the rotor rotates at a speed of 250 rpm. If the whirl velocity upstream from the rotor is zero at all radii, determine

    (a) the axial velocity if the flow onto the stator blade occurs at zero incidence,
    (b) the rotor torque if the axial velocity is constant across the flow annulus, and
    (c) the root and tip rotor blade angles for zero incidence and zero inlet whirl.

2.7 A twelve-bladed axial flow fan has a hydraulic efficiency of 0.92, a mean radius of 0.93 m and rotates at 450 rpm. Air enters the blades axially at a speed of 40 m/s and the head developed across the blades is 35 m of air. If the chord length at the mean radius is 0.33 m, find the blade angles at inlet and outlet if the blades may be considered to act as isolated aerofoils. If the blades are aerofoils with the following characteristics, find also the angle of incidence of the blades and the blade stagger angle. All calculations are to be carried out at the mean radius.

Angle of incidence (deg)	− 4	− 2	0	4	8	12
Lift coefficient	− 0.18	− 0.02	0.13	0.46	0.77	1.025

2.8 A centrifugal pump is used to pump water from a low to a high reservoir having a water level difference of 32 m. The total length of pipe is 1000 m with a friction factor of 0.005 and its diameter is 200 mm. Neglecting all losses except friction, determine the rate of flow between the reservoirs and the power required to drive the pump. The characteristics of the pump are as follows:

Q (m³/h)	0	46	92	138·	184	230
H (m)	68	64	54	42	26.4	8
η (%)	0	49.5	61	63.5	53	10

2.9 The characteristics of a 0.7 m diameter centrifugal pump impeller running at 750 rpm are as follows:

Q (m³/min)	0	7	14	21	28	35	42	49	56
H (m)	40	40.6	40.4	39.3	38.0	33.6	25.6	14.5	0
η (%)	0	41	60	74	83	83	74	51	0

(a) If the pump is initially used to transfer water from one reservoir to another at the same level, determine the pump operating point if the head lost to frictional and other resistances is 35 m at 25 m³/min.

(b) The pump is then used to transfer water between two reservoirs having difference in levels of 15 m through a pipe of 0.45 m diameter. If the pipeline is 93 m long with a friction factor of 0.004 and pipe entry and exit loss coefficients of 0.5 and 1.0 respectively, find the volume flow rate and power absorbed.

(c) If the pump is now changed to one of 0.51 m diameter and the motor is changed to one running at 975 rpm, what is the new volume flow rate and power absorbed?

2.10 A single axial flow water pump has the following characteristics:

Q (m³/s)	0	0.136	0.233	0.311	0.388	0.466	0.608
H (m)	12.6	9.45	9.79	9.07	7.61	5.45	0

A pumping system in which the resistance to flow is purely resistive with no static lift exhibits the same volume flow rate when two of the pumps are connected in parallel as when they are connected in series. What would be the flow rate and head that a single pump would deliver when connected to the same system?

SOLUTIONS

Exercise 2.1 Referring to Fig. 2.5, since there is no slip $\beta_2 = \beta_2^1$.
    The Euler head is given by Eq. (2.2):

$$E = \frac{W}{mg} = \frac{U_2 C_{x2} - U_1 C_{x1}}{g}$$

But $C_{x1} = 0$ since there is no inlet whirl component. Therefore

$$E = \frac{U_2 C_{x2}}{g}$$

$$= \frac{U_2}{g}(U_2 - W_{x2})$$

$$= \frac{10}{9.81}\left(10 - \frac{1.6}{\tan 30°}\right)$$

$$= 7.36 \, \text{m} \quad \text{or} \quad \text{W}/(\text{N/s})$$

Power delivered $= E \times (mg)$

$$= \frac{7.36 \times 10^3 \times 3.5 \times 9.81}{60}$$

$$= 4211.8 \, \text{W}$$

$$\text{Torque delivered} = \frac{\text{Power}}{\text{Angular velocity}}$$

$$= \frac{4211.8 \times 0.65}{10}$$

$$= 273.7 \, \text{N m}$$

Exercise 2.2 Consider first the no-slip condition in Fig. 2.5.

Assuming blades of infinitesimal thickness the flow area may be calculated as

$$\text{Flow area} = \text{Impeller periphery} \times \text{Blade depth}$$

$$= \pi \times 0.02 \times 0.25$$

$$= 15.7 \times 10^{-3} \, \text{m}^2$$

Flow velocity $C_{r2} = Q/A$

$$= \frac{0.028}{15.7 \times 10^{-3}}$$

$$= 1.78 \, \text{m/s}$$

From the outlet velocity triangle

$$W_{x2} = \frac{C_{r2}}{\tan 30°}$$

$$= \frac{1.78}{\tan 30°}$$

$$= 3.08 \, \text{m/s}$$

Now

$$U_2 = \pi DN/60$$

$$= \frac{\pi \times 0.25 \times 1450}{60}$$

$$= 19 \text{ m/s}$$

Absolute whirl component $C_{x2} = U_2 - W_{x2}$

$$= 19 - 3.08$$

$$= 15.92 \text{ m/s}$$

From Eq. (2.2) the Euler head is

$$E = \frac{U_2 C_{x2} - U_1 C_{x1}}{g}$$

and assuming $C_{x1} = 0$ (no whirl at inlet)

$$E = \frac{19 \times 15.92}{9.81}$$

$$= 30.83 \text{ m}$$

From Eq. (2.5)

$$C'_{x2} = \sigma_s \times C_{x2}$$

Therefore the theoretical head with slip is

$$E_N = 0.77 \times 30.83$$

$$= 23.74 \text{ m}$$

The Stodola slip factor is given by Eq. (2.6)

$$\sigma_s = 1 - \frac{\pi \sin \beta_2}{Z[1 - (C_{r2}/U_2)\cot \beta_2]}$$

Then

$$0.77 = 1 - \frac{\pi \sin 30°}{Z[1 - (1.78/19)\cot 30°]}$$

whence

$$Z = 8.15$$

Say, $\underline{\text{number of blades required} = 8}$

Exercise 2.3 (a) Figure 2.5 may be used and in this case $\beta'_2 = \beta_2$ where

$$\beta_2 = 90° - 83°$$

$$= 7°$$

and $\beta_1' = \beta_1$ where

$$\beta_1 = 90° - 75°$$
$$= 15°$$

At inlet, tangential impeller velocity is

$$U_1 = \omega r_1$$
$$= \left(\frac{900 \times 2\pi}{60}\right)(0.051)$$
$$= 4.81 \text{ m/s}$$

From Fig. 2.4

$$\tan \beta_1 = \frac{C_{r1}}{U_1} = \frac{C_1}{U_1} \quad \text{(since zero whirl)}$$
$$C_1 = 4.81 \tan 15°$$
$$= 1.29 \text{ m/s}$$

Volume flow through the pump is

$$Q = A_1 V_1$$
$$= 2\pi r_1 b C_{r1}$$
$$= 2\pi \times 0.051 \times 0.064 \times 1.29$$
$$= \underline{0.0265 \text{ m}^3/\text{s}}$$

(b) Using the continuity equation (Eq. (1.21))

$$r_1 C_{r1} = r_2 C_{r2}$$
$$C_{r2} = \frac{0.051 \times 1.29}{0.203}$$
$$= 0.324 \text{ m/s}$$

At outlet, tangential impeller velocity is

$$U_2 = \omega r_2$$
$$= \left(\frac{900 \times 2\pi}{60}\right)(0.203)$$
$$= 19.13 \text{ m/s}$$

$$\text{Hydraulic efficiency} = \frac{\text{Total head developed by pump}}{\text{Theoretical head developed}}$$

$$\eta_H = \frac{H}{E} \quad \text{(from Eq. (2.22))}$$

If the change in potential head across the pump is ignored, the total head developed by the pump is given by Eq. (2.1):

$$H = \left(\frac{p_2 - p_1}{\rho g}\right) + \left(\frac{C_2^2 - C_1^2}{2g}\right)$$

and for an incompressible fluid the total pressure head difference is

$$\frac{p_{02} - p_{01}}{\rho g} = \left(\frac{p_2}{\rho g} + \frac{C_2^2}{2g}\right) - \left(\frac{p_1}{\rho g} + \frac{C_1^2}{2g}\right) = H$$

Now

$$E = \frac{U_2 C_{x2}}{g} = \frac{U_2}{g}(U_2 - W_{x2})$$

$$= \frac{19.13}{9.81}\left(19.13 - \frac{C_{r2}}{\tan 7^\circ}\right)$$

$$= \frac{19.13}{9.81}\left(19.13 - \frac{0.324}{\tan 7^\circ}\right)$$

$$= 32.15\,\text{m}$$

Therefore

$$H = 0.89 \times 31.91$$
$$= 28.6\,\text{m}$$

whence

$$p_{02} - p_{01} = 28.6 \times 10^3 \times 9.81$$
$$= 278.5\,\text{kPa}$$

At impeller exit

$$C_{x2} = U_2 - \frac{C_{r2}}{\tan \beta_2}$$

$$= 19.13 - \frac{0.324}{\tan 7^\circ}$$

$$= 16.49\,\text{m/s}$$

$$C_2 = (C_{r2}^2 + C_{x2}^2)^{1/2}$$
$$= (0.324^2 + 16.49^2)^{1/2}$$
$$= 16.49\,\text{m/s}$$

Solving for the static pressure head

$$\frac{p_2 - p_1}{\rho g} = H - \left(\frac{C_2^2 - C_1^2}{2g}\right)$$

$$= 28.6 - \left(\frac{16.49^2 - 1.29^2}{2g} \right)$$

$$\underline{p_2 - p_1 = 143.5\,\text{kPa}}$$

(c) Power given to fluid $= \rho g Q H$

$$= 10^3 \times 9.81 \times 0.0265 \times 28.6$$

$$\underline{= 7.13\,\text{kW}}$$

(d) Input power to impeller $= 7.43/0.89$

$$\underline{= 8.35\,\text{kW}}$$

Exercise 2.4 The velocity diagram is shown in Fig. 2.4. From the continuity equation, the velocity in the pipes is

$$v = Q/A$$

$$= \frac{0.04 \times 4}{\pi \times 0.15^2}$$

$$= 2.26\,\text{m/s}$$

Total losses = Pipe friction losses + Other losses

$$= \frac{4flv^2}{2gd} + \frac{3v^2}{2g}$$

$$= \left(\frac{4 \times 0.005 \times 40}{0.15} + 3 \right) \frac{v^2}{2g}$$

$$= \frac{8.333 \times 2.26^2}{2 \times 9.81}$$

$$= 2.16\,\text{m}$$

Total required head $= 35 + 2.16$

$$= 37.16\,\text{m}$$

From Eq. (1.16) the speed of the pump is determined,

$$N_s = \frac{NQ^{1/2}}{(gH)^{3/4}}$$

$$N = \frac{0.075(9.81 \times 37.16)^{3/4}}{(0.04)^{1/2}}$$

$$= 31.28\,\text{rev/s}$$

Flow area perpendicular to impeller outlet periphery

$$= \pi D \times \frac{D}{10} \times 0.94$$
$$= 0.295 \, D^2$$

Now

$$C_{r2} = \frac{Q}{\text{Flow area}}$$
$$= \frac{0.04}{0.295 D^2}$$
$$= \frac{0.136}{D^2} \, \text{m/s}$$

Also

$$U_2 = N\pi D$$
$$= 31.28 \times \pi D$$
$$= 98.3D \, \text{m/s}$$

From Eq. (2.22) the hydraulic efficiency is

$$\eta_{\text{H}} = \frac{\text{Fluid power developed by pump}}{\text{Fluid power supplied to impeller}}$$
$$= \frac{H}{E}$$

or

$$C_{x2} = \frac{gH}{U_2 \eta}$$
$$= \frac{9.81 \times 37.16}{98.3D \times 0.76}$$
$$= 4.87/D \, \text{m/s}$$

The outlet velocity triangle gives

$$\tan 60^\circ = \frac{C_{r2}}{C_{x2} - U_2}$$
$$= \frac{0.136}{D^2(4.87/D - 98.3D)}$$

whence

$$D^3 = 0.0495D - 0.0008$$

Impeller diameter $D = 0.214 \, \text{m}$

Exercise 2.5 Cavitation began when

$$\frac{p_1}{\rho g} + \frac{V_1^2}{2g} = 3.5$$

and at this condition $p_1 = p_{vap}$, the vapour pressure. Therefore

$$\frac{V_1^2}{2g} = 3.5 - \frac{1.8 \times 10^3}{9.81 \times 10^3}$$

$$= 3.317\,\text{m} = NPSH$$

Now

$$\sigma_{TH} = \frac{V_1^2}{2gH}$$

$$= \frac{3.317}{36}$$

$$\sigma_{TH} = 0.0921$$

From the steady flow energy equation (Eq. (2.38)) taking the reservoir level as datum ($Z_0 = 0$) we get for case (1)

$$\frac{p_1}{\rho g} + \frac{V_1^2}{2g} + Z_1 = \frac{p_0}{\rho g} - \text{(Sum of head losses)}$$

$$(Z_1 + h_{f1}) = \frac{p_0}{\rho g} - \sigma_{TH}H - \frac{p_1}{\rho g}$$

$$= (0.75 \times 13.6) - 3.317 - 0.1835$$

$$= 6.7\,\text{m}$$

For case (2)

$$\frac{p_2}{\rho g} + \frac{V_2^2}{2g} + Z_2 = \frac{p_0}{\rho g} - h_{f2}$$

$$(Z_2 + h_{f2}) = (0.62 \times 13.6) - 3.317 - 0.0846$$

$$= 5.03\,\text{m}$$

Since the flow rate is the same, $h_{f1} = h_{f2}$ and the pump must be lowered a distance $(Z_1 - Z_2) = 1.67\,\text{m}$ at the new location.

Exercise 2.6 Referring to the velocity triangles of Fig. 2.24, $i = 0$, $\alpha_2 = 40°$ and $\beta_2 = 30°$ at the mean radius

$$\text{Mean diameter } D_m = \frac{1.8 + 0.75}{2}$$

$$= 1.275\,\text{m}$$

(a) From Eq. (2.43)

$$E = \frac{U}{g}(C_{x2} - C_{x1})$$

$$U = \frac{\pi ND}{60}$$

$$= \frac{250 \times \pi \times 1.275}{60}$$

$$= 16.69 \, \text{m/s}$$

$$W_{x2} = \frac{C_a}{\tan 30°} \quad \text{and} \quad C_{x2} = \frac{C_a}{\tan 40°} \quad \text{(at the mean radius)}$$

Now

$$W_{x2} + C_{x2} = U = C_a\left(\frac{1}{\tan 30°} + \frac{1}{\tan 40°}\right)$$

$$16.69 = 2.92C_a$$

and

$$C_a = 5.71 \, \text{m/s (at the mean radius)}$$

(b) Flow rate Q through the annulus is

$$Q = C_a A$$

$$= \frac{5.71}{4}\pi(1.8^2 - 0.75^2)$$

$$= 12 \, \text{m}^3/\text{s}$$

At the mean radius

$$C_{x2} = \frac{5.71}{\tan 40°}$$

$$= 6.8 \, \text{m/s}$$

$$E = \frac{16.69}{9.81} \times 6.8 \, \text{W/(N/s)}$$

Then power transferred $= \rho g Q E$

$$= \frac{10^3 \times 9.81 \times 12 \times 16.69 \times 6.8}{9.81}$$

$$= 1.362 \, \text{MW}$$

$$\text{Torque} = \frac{\text{Power}}{\text{Angular velocity}}$$

$$= \frac{1.362 \times 10^6 \times 60}{250 \times 2\pi}$$

$$= 52025 \, \text{N m}$$

(c) At the root

$$U_r = \frac{\pi \times 250 \times 0.75}{60}$$

$$= 9.8 \, \text{m/s}$$

Since $C_{x1} = 0$,

$$\tan \beta_{1r} = C_a/U_r$$

$$= 5.71/9.8$$

At the root $\beta_{1r} = 30.2°$

At the tip

$$U_t = \frac{\pi \times 250 \times 1.8}{60}$$

$$= 23.56 \, \text{m/s}$$

$$\tan \beta_{1t} = 5.71/23.56$$

At the tip $\beta_{1t} = 13.6°$

Exercise 2.7 Since the density change across an axial flow fan is so low, it may be considered to be operating with an incompressible fluid and therefore the same equations as apply for axial flow pumps may be used.

$$\text{Hydraulic efficiency} = H/E$$

From Eq. (2.51), $E = 0.5\omega Z C_L W_\infty c/2\pi g$ and putting $\omega = 2UD$ and $Z = \pi D/s$ then

$$E = \frac{C_L \sigma W_\infty U}{2g}$$

Now

$$U = \frac{\pi ND}{60} \qquad \text{at the mean diameter}$$

$$= \frac{\pi \times 450 \times 1.86}{60}$$

$$= 43.83 \, \text{m/s}$$

Also

$$E = UC_{x2}/g$$

and substituting for E

$$\frac{H}{\eta_H} = \frac{UC_{x2}}{g}$$

$$C_{x2} = \frac{35 \times 9.81}{0.92 \times 43.83}$$

$$= 8.51 \text{ m/s}$$

Therefore from Fig. 2.24

$$W_{x2} = U - C_{x1}$$
$$= 43.83 - 8.51$$
$$= 35.32 \text{ m/s}$$

Now

$$\sigma = c/s = cZ/\pi D$$

$$= \frac{0.33 \times 12}{\pi \times 1.86}$$

$$= 0.678$$

But since $C_{x1} = 0$, and therefore $W_{x1} = U$, then

$$W_\infty^2 = C_a^2 + [(W_{x2} + W_{x1})/2]^2$$
$$= 40^2 + [(35.32 + 43.83)/2]^2$$
$$= 3165.78 \text{ m/s}$$
$$W_\infty = 56.26 \text{ m/s}$$

Now

$$C_L W_\infty = \frac{H}{\eta_H} \times \frac{2g}{U} \times \frac{1}{\sigma}$$

$$= \frac{35 \times 2 \times 9.81}{0.92 \times 43.83 \times 0.678}$$

$$= 25.1 \text{ m/s}$$

from which

$$C_L = 25.1/56.26$$
$$= 0.446$$

The aerofoil data are plotted in and the corresponding incidence angle found from Fig. 2.37

$$i = 3.8°$$

Now

$$\sin \beta_\infty = C_a/W_\infty$$
$$\beta_\infty = \sin^{-1}(40/56.26)$$
$$= 45.3°$$

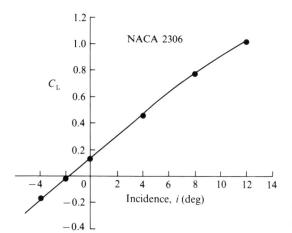

Figure 2.37 Lift coefficient versus incidence angle

$$\tan \beta_1 = C_a/U$$
$$\beta_1 = \tan^{-1}(40/43.83)$$
$$= 42.38°$$

$$\tan \beta_2 = \tan^{-1}(C_a/W_{x2})$$
$$\beta_2 = \tan^{-1}(40/35.32)$$
$$= 48.55°$$

Blade stagger angle $= i + \beta_\infty$
$$= 3.8 + 45.3°$$
$$= 49.1°$$

Exercise 2.8 From Eq. (2.54)

$$\text{System resistance} = H_s + KQ^2$$

Head loss due to pipe friction only is given by

$$h_f = \frac{4flv^2}{2gd}$$

and putting

$$v = Q/A$$

then

$$h_f = \left(\frac{4fl}{2gd}\right)\left(\frac{4}{\pi d^2}\right)^2 Q^2$$

$$= \left(\frac{4 \times 0.005 \times 1000 \times 16}{2 \times 9.81 \times \pi^2 \times 0.2^5}\right)Q^2$$

$$= 5164\, Q^2 \text{m}\, (Q \text{ in m}^3/\text{s})$$

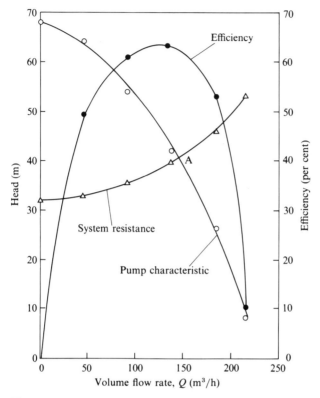

Figure 2.38 Pump and system characteristics

$$\text{System resistance } H = 32 + 5164Q^2 \text{ m}$$

$Q\,(\text{m}^3/\text{h})$	0	46	92	138	184	230
System resistance, $H\,(\text{m})$	32.0	32.8	35.4	39.6	45.5	53.1

The operating point is at the intersection of the pump characteristic and system resistance curves. At the operating point A in Fig. 2.38

$$\underline{Q = 145\,\text{m}^3/\text{h}}$$

$$\underline{H = 40.2\,\text{m}}$$

The efficiency corresponding to the flow rate of $145\,\text{m}^3/\text{h}$ is

$$\eta = 62.5 \text{ per cent}$$

$$\text{Power to drive pump} = \frac{\rho g Q H}{\eta}$$

$$= \frac{10^3 \times 9.81 \times 145 \times 40.2}{60 \times 60 \times 0.625}$$

$$= 25.4\,\text{kW}$$

Exercise 2.9 (a) Figure 2.39 shows the head-flow and efficiency characteristics plotted for the speed of 750 rpm. Since water is being transferred between reservoirs of the same water level, then from Eq. (2.53),

$$\text{System resistance} = KQ^2$$

Solving for K at the point given:

$$K = 35/25^2 = 0.056$$

Therefore the system head loss at the different flow rates may be calculated:

$Q\,(\text{m}^3/\text{min})$	0	7	14	21	28	35	42	49	56
System loss (m)	0	2.74	11	24.7	43.9	68.6	–	–	–

The system resistance curve is now drawn (note that it passes through zero) and the head and flow read off at point A. The corresponding efficiency is read off at point B.

At the operating point

$$Q = 26\,\text{m}^3/\text{min}$$

$$H = 38.3\,\text{m}$$

$$\eta = 81\ \text{per cent}$$

(b) Sum of the head losses and static head is given by Eq. (2.54):

$$H = H_s + KQ^2$$

The head losses may be written as

$$\text{Head losses} = \frac{4flv^2}{2gd} + k_{\text{exit}}\frac{v^2}{2g} + k_{\text{entry}}\frac{v^2}{2g}$$

$$= \left(\frac{4 \times 93 \times 0.004}{0.45} + 1 + 0.5\right)\frac{v^2}{2g}$$

$$= (3.31 + 1 + 0.5)\frac{v^2}{2g}$$

$$= 4.81\,\frac{v^2}{2g}$$

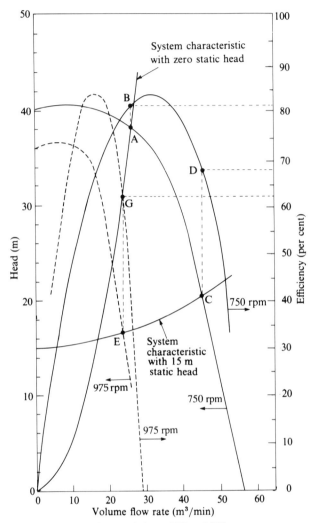

Figure 2.39 Pump characteristics at 750 and 900 rpm

Now

$$v = Q/A$$

and substituting for v

$$\text{Head losses} = \left[\frac{4.81 \times 4^2}{2 \times 9.81 (\pi \times 0.45^2)^2} \right] Q^2$$

$$= 9.69 Q^2 \, \text{m}$$

Including the static lift

$$\text{System loss} = 15 + 9.69Q^2 \,\text{m} \qquad (Q \text{ in } \text{m}^3/\text{s})$$

The head loss is now determined for the various flow rates.

$Q\,(\text{m}^3/\text{min})$	0	7	14	21	28	35	42	49
$H(\text{m})$	15	15.13	15.52	16.18	17.11	18.3	19.75	21.47

The new system resistance curve is drawn noting that it begins at $H = 15\,\text{m}$. The operating point is at point C and the corresponding efficiency at point D.
    At the operating point

$$Q = 45\,\text{m}^3/\text{min}$$

$$H = 20.4\,\text{m}$$

$$\eta = 68.4 \text{ per cent}$$

$$\text{Power absorbed} = \frac{\rho g Q H}{\eta}$$

$$= \frac{10^3 \times 9.81 \times 45 \times 20.4}{0.684 \times 60}$$

$$= 219.4\,\text{kW}$$

(c) Since we have static lift, it is necessary to construct part of the characteristic at the new speed of 900 rpm. The corresponding points for the new impeller and the new speed are found from Eq. (1.6):

$$\frac{Q_1}{N_1 D_1^3} = \frac{Q_2}{N_2 D_2^3} \qquad \text{and} \qquad \frac{H_1}{(N_1 D_1)^2} = \frac{H_2}{(N_2 D_2)^2}$$

whence

$$Q_2 = Q_1 \left(\frac{975}{750}\right)\left(\frac{0.51}{0.7}\right)^3$$

$$= 0.503 Q_1$$

$$H_2 = H_1 \left(\frac{975 \times 0.51}{750 \times 0.7}\right)^2$$

$$= 0.9 H_1$$

Q_1	0	7	14	21	28	35	42	49	56
Q_2	0	3.5	7.1	10.6	14.2	17.7	21.3	24.8	28.3
H_1	40	40.6	40.4	39.3	38	33.6	25.6	14.5	0
H_2	36	36.5	36.4	35.4	34.2	30.24	23.0	13.1	0

The new characteristic is drawn and also the efficiency curve by moving the corresponding values of efficiency horizontally across. The operating point is at E and the corresponding efficiency at G.

At the operating point

$$Q = 23.75 \, \text{m}^3/\text{min}$$

$$H = 16.5 \, \text{m}$$

$$\eta = 62.4 \text{ per cent}$$

$$\text{Power absorbed} = \frac{23.75 \times 16.5 \times 10^3 \times 9.81}{0.624 \times 60}$$

$$= 102.7 \, \text{kW}$$

Exercise 2.10 The single pump characteristic is plotted in Fig. 2.40 along with the characteristics for the pumps connected in parallel and series. Since the same pump is used in both cases, for the series connection the flow rate through the two pumps remains the same while the head is doubled and for the parallel connection the head across the pumps remains the same while the flow rate is doubled.

Series connection

$Q\,(\text{m}^3/\text{s})$	0	0.136	0.233	0.311	0.388	0.466	0.608
$H\,(\text{m})$	25.2	18.9	19.58	18.14	15.22	10.9	0

Parallel connection

$Q\,(\text{m}^3/\text{s})$	0	0.272	0.466	0.622	0.776	0.932	1.216
$H\,(\text{m})$	12.6	9.45	9.79	9.07	7.61	5.45	0

At point A both connections give the same head and flow and the system characteristic must pass through this point and zero since there is no static lift.

At operating point A in Fig. 2.40

$$Q = 0.48 \, \text{m}^3/\text{s}$$
$$H = 9.75 \, \text{m}$$

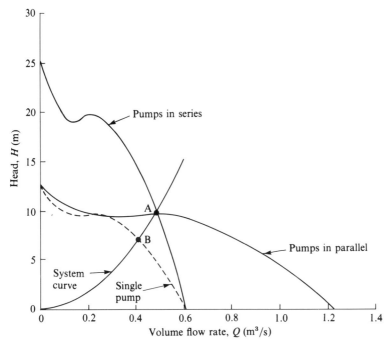

Figure 2.40 Axial flow pump characteristics when connected in series and parallel

From Eq. (2.54) for the system curve

$$H = H_s + KQ^2$$

But $H_s = 0$. Therefore

$$H = 0.48^2 K$$

and

$$K = 42.32$$

The system characteristic is parabolic and may now be drawn in for various heads and flow rates and the point B gives the operating point for the single pump within the system.

System flow rate (m³/s)	0	0.1	0.2	0.3	0.4	0.5	0.6
System head loss (m)	0	0.42	1.69	3.81	6.77	10.58	15.23

The single pump operates at point B:

$$\underline{Q = 0.41 \text{ m}^3/\text{s}}$$

$$\underline{H = 7 \text{ m}}$$

THREE

HYDRAULIC TURBINES

3.1 INTRODUCTION

Turbines are used for converting hydraulic energy into electrical energy. The capital costs of a hydroelectric power scheme (i.e. reservoir, pipelines, turbines, etc.) are higher than thermal stations but they have many advantages, some of which are:

1. High efficiency
2. Operational flexibility
3. Ease of maintenance
4. Low wear and tear
5. Potentially inexhaustible supply of energy
6. No atmospheric pollution

   The main types of turbines used these days are impulse and reaction turbines. The predominant type of impulse machine is the Pelton wheel, which is suitable for a range of heads of about 150–2000 m. One of the largest single units is at the New Colgate Power Station, California, with a rating of 170 MW. Reaction turbines are of two types:

1. Radial or mixed flow
2. Axial flow

   Of the radial flow type the Francis turbine predominates, a single unit at Churchill Falls having a power output of 480 MW with a head of 312 m. Two

Table 3.1 Operating range of hydraulic turbines

	Pelton Wheel	Francis turbine	Kaplan turbine
Dimensionless specific speed (rad)	0.05–0.4	0.4–2.2	1.8–4.6
Head (m)	100–1700	80–500	up to 400
Maximum power (kW)	55	40	30
Best efficiency (%)	93	94	94
Regulation mechanism	Spear nozzle and deflector plate	Guide vanes	Blade stagger

types of axial flow turbines exist, these being the propeller and Kaplan turbines. The former has fixed blades whereas the latter has adjustable blades. Table 3.1 summarizes the head, power and efficiency values that are typical but by no means maxima for each type of turbine.

A reversible pump–turbine can operate as either a pump or a turbine and is used in pump–storage hydroelectric schemes. At times of low electricity demand (e.g. during the night) cheap electricity is used to pump water from the low- to the high-level reservoir. This water may then be used during the day for power generation during peak periods, when the unit runs as a turbine in the reverse direction.

One of the largest pump–storage schemes in the world is at Cabin Creek in Colorado, where each turbine generates 166 MW with a head of 360 m.

In the sections that follow, each type of hydraulic turbine will be studied separately in terms of the velocity triangles, efficiencies, reaction and method of operation.

3.2 PELTON WHEEL

The Pelton wheel turbine is a pure impulse turbine in which a jet of fluid issuing from a nozzle impinges on a succession of curved buckets fixed to the periphery of a rotating wheel, as in Fig. 3.1, where four jets are shown. The buckets deflect the jet through an angle of between 160 and 165° in the same plane as the jet, and it is the turning of the jet that causes the momentum change of the fluid and its reaction on the buckets. A bucket is therefore pushed away by the jet and the next bucket moves round to be similarly acted upon. The spent water falls vertically into the lower reservoir or tailrace and the whole energy transfer from nozzle outlet to tailrace takes place at constant pressure. Figure 3.2 shows a large Pelton wheel with its buckets.

A diagram of a Pelton wheel hydroelectric installation is shown in Fig. 3.3. The water supply is from a constant-head reservoir at elevation H_1

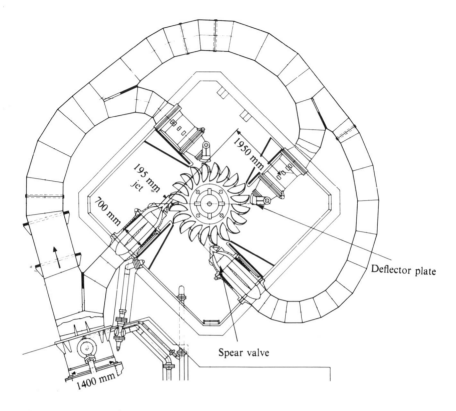

Figure 3.1 Elements of a Pelton wheel turbine (*courtesy of Escher Wyss Ltd*)

above the centre-line of the jet. A shallow-slope pressure tunnel extends from the reservoir to a point almost vertically above the location of the turbine. A pipe of almost vertical slope called the penstock joins the end of the pressure tunnel to the nozzle, while a surge tank is installed at the upper end of the penstock to damp out flow control pressure and velocity transients. It is emphasized that, compared with the penstock, the pressure tunnel could be extremely long, its slope is extremely shallow and it should undergo no large pressure fluctuations caused by inlet valve flow control. The penstock must be protected against the large pressure fluctuations that could occur between the nozzle and surge tank, and is usually a single steel-lined concrete pipe or a steel-lined excavated tunnel. At the turbine end of the penstock is the nozzle, which converts the total head at inlet to the nozzle into a water jet with velocity C_1 at atmospheric pressure.

The velocity triangles for the flow of fluid onto and off a single bucket are shown in Fig. 3.4. If the bucket is brought to rest, then subtracting the bucket speed U_1 from the jet velocity C_1 gives the relative fluid velocity W_1 onto the bucket. The angle turned through by the jet in the horizontal plane during its passage over the bucket surface is α and the relative exit velocity is W_2. If the

Figure 3.2 Pelton wheel (*courtesy of Escher Wyss Ltd*)

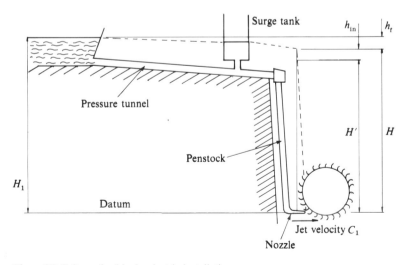

Figure 3.3 Pelton wheel hydroelectric installation

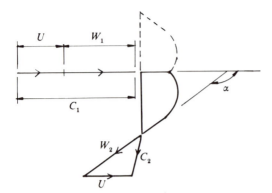

Figure 3.4 Velocity triangles for a Pelton wheel

bucket speed vector U_2 is added to W_2 in the appropriate direction, the absolute velocity at exit, C_2, results. It should be realized that the component C_{x2} of C_2 can be in the positive or negative x direction depending on the magnitude of U.

From Euler's turbine equation (Eq. (1.24))

$$W/m = U_1 C_{x1} - U_2 C_{x2}$$

and since in this case C_{x2} is in the negative x direction,

$$W/m = U\{(U + W_1) + [W_2 \cos(180° - \alpha) - U]\}$$

Assuming no loss of relative velocity due to friction across the bucket surface ($W_1 = W_2$), then

$$W/m = U(W_1 - W_1 \cos \alpha)$$

Therefore

$$E = U(C_1 - U)(1 - \cos \alpha)/g \qquad (3.1)$$

the units of E being watts per newton per second weight of flow.

Equation (3.1) can be optimized by differentiating with respect to U. Thus

$$dE/dU = (1 - \cos \alpha)(C_1 - 2U)/g = 0$$

for a maximum, and then

$$C_1 = 2U$$

or

$$U = C_1/2 \qquad (3.2)$$

Substituting back into Eq. (3.1) gives

$$E_{max} = C_1^2(1 - \cos \alpha)/4g \qquad (3.3)$$

In practice, surface friction of the bucket is present and $W_2 \neq W_1$. Then

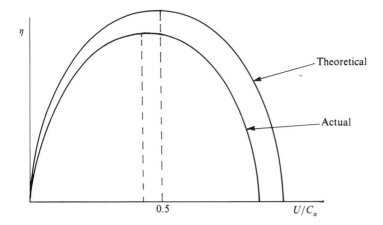

Figure 3.5 Efficiencies and jet speed ratio of a Pelton wheel

Eq. (3.1) becomes

$$E = U(C_1 - U)(1 - k\cos\alpha)/g$$

where k is the relative velocity ratio W_2/W_1.

If the hydraulic efficiency is defined as

$$\eta_H = \frac{\text{Energy transferred}}{\text{Energy available in jet}}$$

$$= E/(C_1^2/2g) \tag{3.4}$$

then if $\alpha = 180°$, the maximum hydraulic efficiency is 100 per cent. In practice, the deflection angle is in the order of 160–165° to avoid interference with the oncoming jet and η_H is accordingly reduced. Figure 3.5 shows the theoretical efficiency as a function of speed ratio. The overall efficiency is lower than the theoretical as well as having a reduced speed ratio at maximum efficiency. This is due to pipeline and nozzle losses, which will be discussed in a later section.

3.2.1 Pelton Wheel Load Changes

Hydraulic turbines are usually coupled directly to an electrical generator and, since the generator must run at a constant speed, the speed U of the turbine must remain constant when the load changes. It is also desirable to run at maximum efficiency and therefore the ratio U/C_1 must stay the same. That is, the jet velocity must not change. The only way left to adjust to a change in turbine load is to change the input water power.

The input water power is given by the product $\rho g Q H'$ but H' is constant

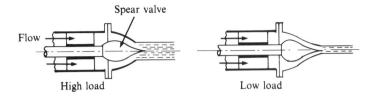

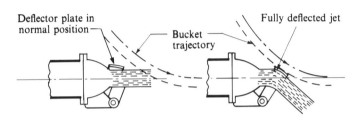

Figure 3.6 Load control by a spear valve and deflector plate

(and therefore C_1), and the only remaining variable is Q. The change in flow rate is effected by noting that $Q = C_1 A$, where A is the nozzle area. Since C_1 is constant, then the cross-sectional area of the nozzle must change. This is accomplished by a spear valve, which alters the jet cross-sectional area as illustrated in Fig. 3.6. The position of the spear is controlled by a servo-mechanism that senses the load change. For a sudden loss of load, a deflector plate rises to remove the jet totally from the buckets and to allow time for the spear to move slowly to its new load position. This prevents excessive overspeeding.

3.2.2 Pelton Wheel Losses and Efficiencies

Head losses occur in the pipelines conveying the water to the nozzle and are composed of friction and bend losses. Losses also occur in the nozzle and these are expressed in terms of a velocity coefficient C_v. Finally there are windage and friction losses in the wheel itself. The total head line is shown in Fig. 3.3, where the water supply is from a reservoir at a head H_1 above the nozzle. As the fluid moves through the pressure tunnel and penstock up to entry to the nozzle, a frictional head loss h_f occurs. A further head loss h_{in} due to losses in the nozzle takes place so that the head available for power generation at exit from the nozzle is H':

$$H' = H_1 - (h_f + h_{in})$$
$$= C_1^2/2g$$

when expressed as kinetic energy per unit weight of flow. Now

$$\text{Pipeline transmission efficiency} = \frac{\text{Energy at end of pipeline}}{\text{Energy available at reservoir}}$$

or

$$\eta_{\text{trans}} = (H_1 - h_f)/H_1 = H/H_1 \qquad (3.5)$$

and

$$\text{Nozzle efficiency} = \frac{\text{Energy at nozzle outlet}}{\text{Energy at nozzle inlet}}$$

or

$$\eta_N = H'/(H_1 - h_f) = C_1^2/2gH \qquad (3.6)$$

So

$$\text{Nozzle and pipe transmission efficiency} = (H/H_1)(H'/H) = C_1^2/2gH_1$$

Also

$$\text{Nozzle velocity coefficient} = \frac{\text{Actual jet velocity}}{\text{Theoretical jet velocity}}$$

or

$$C_v = C_1/(2gH)^{1/2} \qquad (3.7)$$

Therefore the nozzle efficiency becomes

$$\eta_N = C_1^2/2gH = C_v^2 \qquad (3.8)$$

The characteristics of an impulse turbine are shown in Figs 3.7 and 3.8. These curves are drawn for a constant head and it is seen that peak power and efficiency occur at about the same speed ratio for any gate opening and that the peak values of efficiency do not vary much. This is due to the nozzle velocity remaining constant in magnitude and direction as the flow rate changes, giving an optimum value of U/C_1 at a fixed speed. Windage,

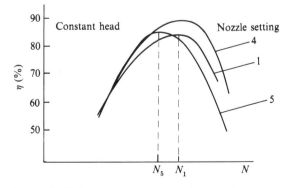

Figure 3.7 Efficiency versus speed at various nozzle settings

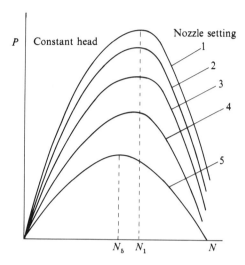

Figure 3.8 Pelton wheel power output versus speed at various nozzle settings

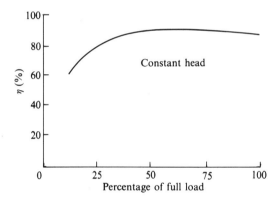

Figure 3.9 Variation of Pelton wheel efficiency with load

mechanical losses and variations in loss coefficients cause the small variations. In practice, one is usually more interested in the fixed-speed condition since the generators run at constant speed. Figure 3.9 shows that the variation of efficiency with load is slight, except at low loads where the decrease is due to changes in nozzle efficiency, and at high loads where the increased jet diameter gives rise to higher bucket losses.

3.3 RADIAL FLOW TURBINE

The radial flow or Francis turbine is a reaction machine and, to achieve reaction, the rotor must be enclosed by a casing to prevent deviation of the fluid streamlines around the edge of the blades. The essential difference between the reaction rotor and impulse wheel is that in the former the water, under a high static head, has its pressure energy transformed into kinetic energy in a nozzle, which in itself forms part of the rotor. Therefore, since a static pressure drop occurs across the turbine rotor, part of the work done by the fluid on the runner is due to reaction from the pressure drop, and part is due to a change in kinetic energy, which represents an impulse function. Figure 3.10 shows a cross section through a Francis turbine. A typical rotor is also shown.

The total head of the radial flow turbine ranges from about 500 down to 30 m, and the dimensionless specific speed from 0.3 to 2.5 rad. The water first enters a volute or spiral casing. It then passes through a row of fixed guide vanes followed by adjustable guide vanes, the cross-sectional area between the adjustable vanes being varied for flow control at part load. The water then passes immediately into the rotor where it moves radially through the rotor vanes and exits from the rotor blades at a smaller diameter, after which it turns through 90° into the draft tube. The function of the draft tube is to bring the water pressure back to the pressure of the tailrace, and during this process to remove the kinetic energy still existing at the runner outlet. In some rotors the work transfer is accomplished not only while the water is moving radially but also in a part axial direction. This is done by a judicious choice of rotor design.

In considering the flow through the runner of a radial flow turbine and the theoretical analysis with respect to the velocity triangles, the same assumptions will be made as for the centrifugal pump. Figure 3.11 shows the total energy line through the complete system from supply reservoir to tailrace. The free surface of the reservoir is often called the 'head water'. The fluid experiences a frictional head drop h_{fp} in the pipeline up to the inlet flange of the turbine at point 0. At this point the fluid enters the volute, the fixed inlet guide vanes and also the adjustable guide vanes, where a further frictional head loss h_g is experienced. Upon leaving the inlet guide vanes, the fluid moves into the runner where a frictional head loss h_r takes place while energy is supplied to the runner and shaft. The total energy line decreases to the point 3 where the fluid exits from the draft tube with a frictional head loss h_d occurring in the draft tube, and a residual kinetic energy loss $V_3^2/2g$ at exit from the draft tube. In considering the overall turbine efficiency, the inlet volute and draft tube are considered to be parts of the turbine.

The inlet and outlet velocity triangles for the runner are shown in Fig. 3.12. This shows the water emerging from the inlet guide vanes at radius r_1 with absolute velocity C_1, at an angle α_1 to the direction of rotation. The tangential velocity at inlet is U_1, and by subtracting U_1 from C_1 the relative velocity vector W_1 is obtained at an angle β_1 to the direction of rotation. β_1 is also the inlet blade angle for shock-free entry. At the outlet radius r_2,

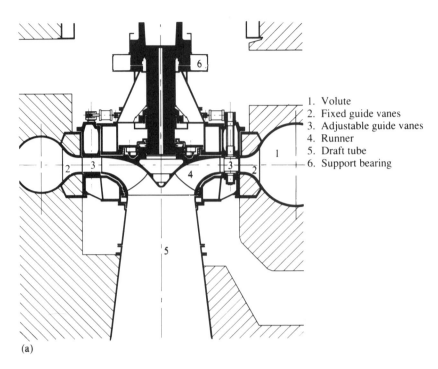

(a)

1. Volute
2. Fixed guide vanes
3. Adjustable guide vanes
4. Runner
5. Draft tube
6. Support bearing

Figure 3.10 A Francis turbine (*courtesy of Escher Wyss Ltd*): (a) hydroelectric installation; (b) turbine runner

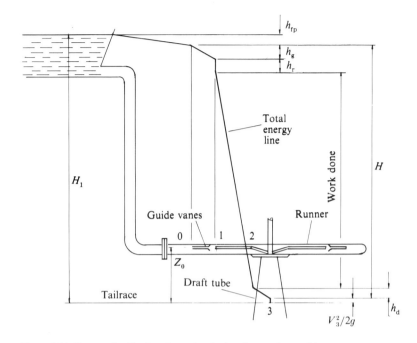

Figure 3.11 Energy distribution through a hydraulic reaction turbine

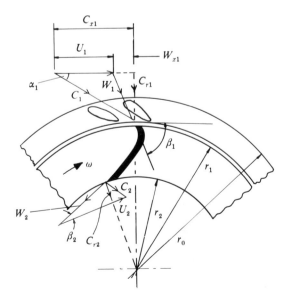

Figure 3.12 Velocity triangles for a Francis turbine

99

the water leaves the blade at angle β_2 to the tangential velocity vector, the resultant of W_2 and U_2 being the absolute outlet velocity C_2. The flow velocities C_{r1} and C_{r2} are directed towards the axis of rotation and are given by $Q/2\pi r_1 b_1$ and $Q/2\pi r_2 b_2$ respectively, where b is the height of the runner.

Euler's turbine equation (Eq. (1.24)) gives

$$E = W/mg = (U_1 C_{x1} - U_2 C_{x2})/g \quad (\text{J/N}) \qquad (3.9)$$

and E is a maximum when C_{x2} is zero, that is when the absolute and flow velocities are equal at the outlet.

3.3.1 Net Head Across Reaction Turbine

The net head H across the turbine is the difference in the total head between the inlet flange and the tail water level. Gross head H_1 should not be confused with net head. Thus

$$\text{Turbine total inlet head} = p_0/\rho g + V_0^2/2g + Z_0$$

and

$$\text{Turbine total outlet head} = p_3/\rho g + V_3^2/2g + Z_3$$

where the pressures are gauge pressures. Summing these,

$$\text{Total head across turbine} = (p_0 - p_3)\rho g + (V_0^2 - V_3^2)/2g + (Z_0 - Z_3)$$
$$= H$$

But in the tailrace p_3 is atmospheric and Z_3 is zero. Therefore

$$H = (p_0/\rho g + V_0^2/2g + Z_0) - V_3^2/2g \qquad (3.10)$$

Also

$$H = H_1 - h_{fp} - V_3^2/2g \qquad (3.11)$$

and the energy given to the runner by the water per unit weight of flow is

$$W/mg = H - h_d - h_g - h_r \qquad (3.12)$$

If the water discharged directly into the tailrace from the runner outlet, the kinetic energy lost would be high. By fitting the draft tube between runner outlet and tailrace, a continuous stream of water is formed between the two. The tailrace velocity is reduced because of the increase in cross-sectional area of the draft tube, and, because the tailrace pressure is atmospheric, the runner outlet pressure must now be below atmospheric pressure. Applying the energy equation between the runner outlet and tailrace gives

$$p_2/\rho g + V_2^2/2g + Z_2 = p_3/\rho g + V_3^2/2g + Z_3 + h_d$$

and putting p_3 and Z_3 equal to zero

$$p_2/\rho g = (V_3^2 - V_2^2)/2g - Z_2 + h_d \qquad (3.13)$$

There is a limit to the amount that V_3 can be reduced because of tube length. As the length increases, h_d also increases, and since the angle of divergence should not exceed about $8°$, to ensure that separation of the boundary layer does not occur, the draft tube could be very long. There is also an upper limit on the value of Z_2 due to cavitation, and this is discussed in a later section.

3.3.2 Turbine Losses

The losses can once again be related in terms of an energy balance through the turbine:

$$P = P_m + P_r + P_c + P_l + P_s \qquad (3.14)$$

where P_s = shaft power output, P_m = mechanical power loss, P_r = runner power loss, P_c = casing and draft tube loss, P_l = leakage loss and P = water power available. Together, $P_r + P_c + P_l$ is the hydraulic power loss.

Runner power loss P_r is due to friction, shock at impeller entry and flow separation, and results in a head loss h_r associated with a flow rate through the runner of Q_r:

$$P_r = \rho g Q_r h_r \ (\text{N m/s}) \qquad (3.15)$$

Leakage power loss P_l is caused by a flow rate q leaking past the runner and therefore not being handled by the runner. Thus

$$Q = Q_r + q \qquad (3.16)$$

and, with a total head H_r across the runner, the leakage power loss becomes

$$P_l = \rho g H_r q \ (\text{N m/s}) \qquad (3.17)$$

Casing power loss P_c is due to friction, eddy and flow separation losses in the casing and draft tube. If this head loss is h_c then

$$P_c = \rho g Q h_c \ (\text{N m/s}) \qquad (3.18)$$

The total energy balance of Eq. (3.14) thus becomes

$$\rho g Q H = P_m + \rho g (h_r Q_r + h_c Q + H_r q + P_s) \qquad (3.19)$$

Then we have

$$\text{Overall efficiency} = \frac{\text{Shaft output power}}{\text{Fluid power available at inlet flange}}$$

or

$$\eta_0 = P_s / \rho g Q H \qquad (3.20)$$

and

$$\text{Hydraulic efficiency} = \frac{\text{Power received by runner}}{\text{Fluid power available at inlet flange}}$$

or

$$\eta_H = (P_s + P_m)/\rho g Q H \qquad (3.21)$$

The term $(P_s + P_m)/\rho g Q H$ of Eq. (3.21) is the theoretical energy transfer per unit weight of fluid flow. Therefore for maximum efficiency

$$\eta_H = U_1 C_{x1}/gH$$

3.3.3 Characteristic Curves

Curves of water power input, torque exerted by the wheel, flow rate, efficiency and brake power output for a constant gate opening are drawn in Fig. 3.13.

In comparison with the impulse wheel, the flow rate is no longer independent of the wheel speed since there is now an unbroken flow of water down to the tailrace and changes within the runner affect the flow. Of most interest, however, is the behaviour of the turbine at constant speed, since the generator runs at a fixed speed. As the electrical load changes, so the flow rate is changed by variation of the gate opening. From a set of characteristic curves at different gate openings, the constant-speed characteristic of Fig. 3.14 is constructed. It is seen that the head increases slightly as the load is decreased due to the friction head loss, which, being proportional to Q^2, is less at lighter loads. It will also be noted that the efficiency curve at constant speed is not as flat as for an impulse turbine.

When a reaction turbine operating at constant speed experiences a load decrease, the cross-sectional area between the inlet vanes changes and angle α_1

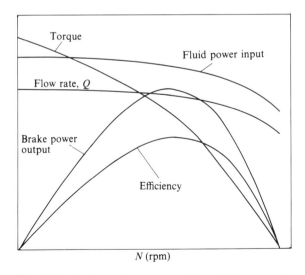

Figure 3.13 Reaction turbine characteristics at full gate opening

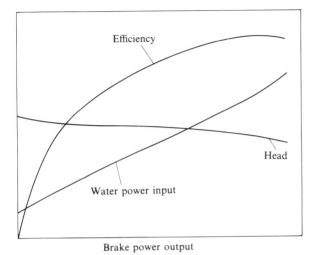

Brake power output

Figure 3.14 Reaction turbine characteristics at constant speed but varying gate opening

decreases. The inlet runner area is constant and therefore to satisfy continuity the relative velocity W_1 must decrease. The result is that the flow onto the runner is no longer shock-free and at exit C_2 may increase. This gives a higher kinetic energy loss at runner exit as well as increasing the whirl component C_{x2} down the draft tube. The flow is then spiral in nature, which decreases the draft tube efficiency. The efficiency of a reaction turbine at light loads therefore tends to be less than that of the Pelton wheel, although the design maximum efficiency may be greater.

3.4 AXIAL FLOW TURBINE

The energy distribution through an axial flow hydraulic turbine is the same as in Fig. 3.11. However, the velocity triangles are markedly different since the fluid is assumed to flow from blade inlet to outlet at a constant radius. A Kaplan turbine is illustrated in Fig. 3.15. The inlet guide vanes are fixed and are situated at a plane higher than the runner blades such that the fluid must turn through 90° to enter the runner in the axial direction. Load changes are effected by adjustment of the runner blade angle. The function of the guide vanes is to impart whirl to the fluid so that the radial distribution of velocity is the same as in a free vortex. Since this type of turbine is used for low heads and high flow rates, the blades must be long and have large chords so that they are strong enough to transmit the very high torques that arise. Pitch/chord ratios of 1–1.5 are typical for axial flow turbines and this results in a four-, five- or six-bladed runner.

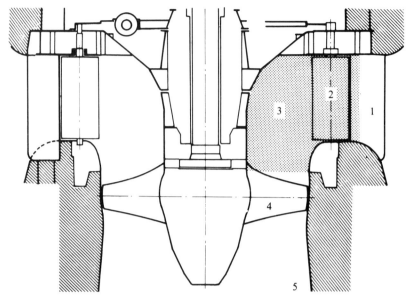

1. Spiral casing with fixed guide vanes
2. Adjustable inlet guide vanes
3. Transition passage
4. Runner
5. Draft tube

Figure 3.15 Axial flow Kaplan turbine (*courtesy of Escher Wyss Ltd*)

The velocity triangles are usually drawn at the mean radius, since conditions change from hub to tip, and are shown in Fig. 3.16. The flow velocity is axial at inlet and outlet and hence $C_{r1} = C_{r2} = C_a$. The blade velocity vector U_1 is subtracted from the absolute velocity vector C_1 (which is at angle α_1 to U_1) to yield the relative velocity vector W_1. For shock-free entry onto the runner, W_1 is at the blade angle β_1. For maximum efficiency, the whirl component C_{x2} is zero, in which case the absolute velocity at exit is axial, and then $C_2 = C_{r2}$. Euler's turbine equation (Eq. (1.24)) gives

$$E = U(C_{x1} - C_{x2})/g \,(\text{J/N})$$

and for zero whirl at exit

$$E = UC_{x1}/g \tag{3.22}$$

Now

$$C_{x1} = U - C_a \cot(180° - \beta_1) = U + C_a \cot \beta_1$$

and therefore

$$E = (U^2 + UC_a \cot \beta_1)/g \tag{3.23}$$

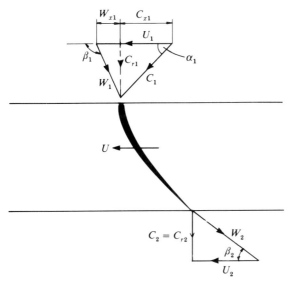

Figure 3.16 Velocity triangles for an axial flow hydraulic turbine

If E is constant along the blade radius, and C_a is constant over the cross-sectional area, then as U^2 increases from hub to tip, $U \cot \beta_1$ must decrease to keep Eq. (3.23) constant. That is, β_1 must increase from hub to tip and the blade must therefore be twisted.

The characteristic curves for the axial flow Kaplan turbine are similar to those of the radial flow turbine, and a comparison of the efficiencies of

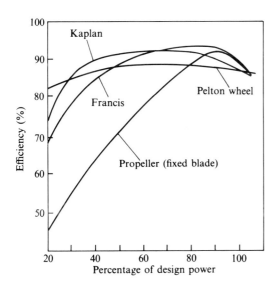

Figure 3.17 Comparison of hydraulic turbine efficiencies

the impulse and two reaction wheels at part load is shown in Fig. 3.17. Of the three, the impulse wheel efficiency curve is much flatter although the maximum efficiency is lower. The Francis turbine peaks at the highest efficiency but falls off rapidly at part load, while the Kaplan turbine has a much flatter curve than the Francis and exhibits a similar maximum efficiency. The advantage of adjustable blades on the Kaplan turbine is shown by comparing it with the curve for a fixed-blade propeller turbine (also shown in Fig. 3.17).

3.5 CAVITATION IN TURBINES

Turbine cavitation occurs on the suction surfaces of the blades, at the runner outlet, where the static pressure is a minimum and the absolute velocity high. Although it has little if any effect on the performance of the turbine since it occurs after the runner, it should be avoided if at all possible. Referring again to Eq. (3.13), as the outlet velocity V_2 increases then p_2 decreases and has its lowest value when the vapour pressure is reached. At this pressure cavitation begins, and putting p_3 equal to p_{atm} and p_2 equal to p_{vap}, Eq. (3.13) becomes

$$[(V_2^2 - V_3^2)/2g] - h_d = (p_{atm} - p_{vap})/\rho g - Z_2 \qquad (3.24)$$

Division of Eq. (3.24) by the net head across the turbine gives the Thoma cavitation parameter for the turbine:

$$\sigma = [(p_{atm} - p_{vap})/\rho g - Z_2]/H = (NPSH)/H \qquad (3.25)$$

The critical value of $NPSH$ at which cavitation occurs is determined from a test on a model or full-size machine in which p_2 is decreased until the minimum value at which cavitation begins or the efficiency suddenly decreases is found. Knowing Z_2 and H it is easy to compute the critical value σ_c, which is the value below which σ, as given by Eq. (3.25), for any other similar machine of the same homologous series must not fall.

Equation (3.25) shows that the maximum elevation of the turbine above the tailrace is given by

$$Z_2 = (p_{atm} - p_{vap})/\rho g - \sigma_c H \qquad (3.26)$$

Equation (3.26) indicates that, as the net head is increased, so the turbine elevation above the tailrace must be decreased. For an excessive net head, Z_2 might be negative, which implies that excavation would be needed to place the turbine below the level of the tailrace.

The dependence of σ_c on the dimensionless specific speed and thus on the design of the turbine is shown in Fig. 3.18 where turbines of high N_s have a high σ_c and must therefore be set lower than those of smaller N_s. The similarity relationships used for pumps in connection with cavitation may also be used for turbines.

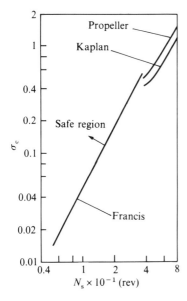

Figure 3.18 Cavitation limits for reaction turbines

EXERCISES

3.1 A generator is to be driven by a small Pelton wheel with a head of 91.5 m at inlet to the nozzle and discharge of 0.04 m³/s. The wheel rotates at 720 rpm and the velocity coefficient of the nozzle is 0.98. If the efficiency of the wheel (based on the energy available at entry to the nozzle) is 80 per cent and the ratio of bucket speed to jet speed is 0.46, determine the wheel-to-jet-diameter ratio at the centre-line of the buckets, and the speed of the wheel. What is the dimensionless power specific speed of the wheel?

3.2 A reservoir with a height of 280 m is connected to the powerhouse of a hydroelectric plant through three pipes each 2.5 km long and with friction factor 0.006, in which the head loss is not to exceed 34 m. It is a requirement that a total shaft output of 18 MW be developed, and to achieve this it is decided to install a number of single-jet Pelton wheels, each with a dimensionless specific speed not exceeding 0.23 rad. The ratio of bucket speed to jet speed is 0.46, while the wheel speed is to be 650 rpm. If the nozzles have a discharge coefficient of 0.94 and velocity coefficient of 0.96, and assuming that each wheel has an overall efficiency of 87 per cent, find
    (a) the number of Pelton wheels required,
    (b) the wheel diameter,
    (c) the jet nozzle diameter and
    (d) the diameter of the supply pipes.

3.3 The buckets of a Pelton wheel deflect the jet through an angle of 170°, while the relative velocity of the water is reduced by 12 per cent due to bucket friction. Calculate the theoretical hydraulic efficiency from the velocity triangles for a bucket/jet speed ratio of 0.47. Under a gross head of 600 m the wheel develops 1250 kW when the loss of head due to pipe friction between the reservoir and nozzle is 48 m. The bucket circle diameter of the wheel is

900 mm and there are two jets. The nozzle velocity coefficient is 0.98. Find the speed of rotation of the wheel and the diameter of the nozzles if the actual hydraulic efficiency is 0.9 times that calculated above.

3.4 An electricity generating installation uses a Francis turbine with a rotational speed of 1260 rpm. The net head across the turbine is 124 m and the volume flow rate is 0.5 m³/s. The radius of the runner is 0.6 m, the height of the runner vanes at inlet is 0.03 m and the angle of the inlet guide vanes is set at 72° from the radial direction. Assuming that the absolute flow velocity is radial at exit, find the torque and power exerted by the water. Calculate the hydraulic efficiency.

3.5 An inward flow radial turbine has an overall efficiency of 74 per cent. The net head H across the turbine is 5.5 m and the required power output is 125 kW. The runner tangential velocity is $0.97(2gH)^{1/2}$ while the flow velocity is $0.4(2gH)^{1/2}$. If the speed of the runner is 230 rpm with hydraulic losses accounting for 18 per cent of the energy available, calculate the inlet guide vane exit angle, the inlet angle of the runner vane, the runner diameter at inlet and the height of the runner at inlet. Assume that the discharge is radial.

3.6 A Francis turbine has a diameter of 1.4 m and rotates at 430 rpm. Water enters the runner without shock with a flow velocity (C_{r1}) of 9.5 m/s and leaves the runner without whirl with an absolute velocity of 7 m/s. The difference between the sum of the static and potential heads at entrance to the runner and at the exit from the runner is 62 m. If the turbine develops 12 250 kW and has a flow rate of 12 m³/s of water when the net head is 115 m, find

  (a) the absolute velocity of the water at entry to the runner and the angle of the inlet guide vanes,
  (b) the entry angle of the runner blades and
  (c) the head lost in the runner.

3.7 An inward flow vertical shaft reaction turbine runs at a speed of 375 rpm under an available net total head from inlet flange to tailrace of 62 m. The external diameter of the runner is 1.5 m and the dimensionless power specific speed based on the power transferred to the runner is 0.14 rev. Water enters the runner without shock with a flow velocity of 9 m/s and leaves the runner without whirl with an absolute velocity of 7 m/s. It discharges to the tailrace with a velocity of 2.0 m/s. The mean height of the runner entry plane is 2 m above the tailrace level while the entrance to the draft tube is 1.7 m above the tailrace. At entrance to the runner the static pressure head is 35 m above atmospheric pressure, while at exit from the runner the static pressure head is 2.2 m below atmospheric pressure. Assuming a hydraulic efficiency of 90 per cent, find

  (a) the runner blade entry angle,
  (b) the head lost in the volute casing and guide vanes, in the runner and in the draft tube and
  (c) the draft tube entry diameter.

3.8 An axial flow hydraulic turbine has a net head of 23 m across it, and, when running at a speed of 150 rpm, develops 23 MW. The blade tip and hub diameters are 4.75 and 2.0 m respectively. If the hydraulic efficiency is 93 per cent and the overall efficiency 85 per cent, calculate the inlet and outlet blade angles at the mean radius assuming axial flow at outlet.

3.9 A Kaplan turbine operating under a net head of 20 m develops 16 000 kW with an overall efficiency of 80 per cent. The diameter of the runner is 4.2 m while the hub diameter is 2 m and the dimensionless power specific speed is 3 rad. If the hydraulic efficiency is 90 per cent, calculate the inlet and exit angles of the runner blades at the tip and at the hub if the flow leaving the runner is purely axial.

3.10 An axial flow turbine with tip and hub diameters of 2.0 and 0.8 m respectively rotates at 250 rpm. The turbine is fitted with fixed stator blades upstream of the rotor and at the mean diameter these are set at 42° to the direction of blade rotation. Also, at the mean diameter and measured from the direction of blade rotation, the blade angle at inlet is 148°.

  (a) Assuming a constant axial velocity across the annulus, what is the flow rate for which the angle of incidence of the rotor blades is zero?

(b) Determine the outlet angle of the rotor blade if the whirl component at outlet is zero.

(c) Calculate the theoretical power output if the whirl at inlet and outlet is the same at all radii.

3.11 A hydraulic turbine is designed to run at 300 rpm under a net head of 50 m and to produce 2 MW of power. The runner outlet velocity of the fluid is expected to be 10.4 m/s and it is proposed to situate the runner outlet at a height of 6 m above the tailrace. The atmospheric pressure is equivalent to 10.3 m of water and the saturation pressure of water is 4 kPa. Determine

(a) whether cavitation is likely to occur,

(b) the limiting height of the runner outlet if cavitation is to be just avoided and

(c) the head loss between runner outlet and tailrace when cavitation is just about to occur.

Critical cavitation parameters are given below:

N_s (rev)	0	0.049	0.096	0.144	0.192	0.24
σ_c	0	0.04	0.1	0.18	0.28	0.41

SOLUTIONS

Exercise 3.1 Overall efficiency

$$\eta_0 = \frac{\text{Power developed}}{\text{Power available}}$$

$$P = \rho g Q H \eta_0$$
$$= 10^3 \times 9.81 \times 0.04 \times 91.5 \times 0.8$$

Power developed $= 28.72\,\text{kW}$

Velocity coefficient

$$C_v = \frac{C_1}{(2gH)^{1/2}}$$

$$C_1 = 0.98(2 \times 9.81 \times 91.5)^{1/2}$$
$$= 41.52\,\text{m/s}$$

Therefore

$$U = 0.46C_1$$
$$= 0.46 \times 41.52$$
$$= 19.1\,\text{m/s}$$

Also

$$U = \frac{\omega D}{2} \quad \text{where } D \text{ is wheel diameter}$$

$$D = \frac{2 \times 19.1 \times 60}{720 \times 2\pi}$$

$$= 0.507\,\text{m}$$

Jet area

$$A = Q/C_1$$

$$= \frac{0.04}{41.52}$$

$$= 0.963 \times 10^{-3} \, \text{m}^2$$

and jet diameter

$$d = \left(\frac{4A}{\pi}\right)^{1/2}$$

$$= \left(\frac{4 \times 0.963 \times 10^{-3}}{\pi}\right)^{1/2}$$

$$= 0.035 \, \text{m}$$

Diameter ratio

$$\frac{D}{d} = \frac{0.507}{0.035}$$

$$= 14.5$$

Dimensionless power specific speed is given by Eq. (1.20):

$$N_{sp} = \frac{NP^{1/2}}{\rho^{1/2}(gH)^{5/4}}$$

$$= \frac{720}{60} \times \left(\frac{28.72 \times 10^3}{10^3}\right)^{1/2} \times \left(\frac{1}{9.81 \times 91.5}\right)^{5/4}$$

$$= 0.0131 \, \text{rev}$$

$$= 0.0131 \times 2\pi \, \text{rad}$$

$$\underline{\text{Power specific speed} = 0.082 \, \text{rad}}$$

A check with Fig. 1.10 shows this value of N_{sp} to be within the range for a Pelton wheel.

Exercise 3.2 (a) From Eq. (1.20) we get the power specific speed for each wheel

$$N_{sp} = \frac{NP^{1/2}}{\rho^{1/2}(gH)^{5/4}} \qquad \text{where } N \text{ is in rpm}$$

Available head

$$H = \text{Gross head} - \text{Head loss}$$

$$= 280 - 34$$

$$= 246 \, \text{m}$$

Available power per wheel

$$P = \left(\frac{0.23 \times 60}{2\pi \times 650}\right)^2 (9.81 \times 246)^{5/2} \times 1000$$

$$= 3266 \, \text{kW}$$

$$\text{Number of wheels} = \frac{\text{Total power output}}{\text{Power per machine}}$$

$$= \frac{18 \times 10^6}{3.266 \times 10^6}$$

$$= 5.51 \, (\text{say 6 machines})$$

(b) Nozzle velocity coefficient

$$C_v = \frac{C_1}{(2gH)^{1/2}}$$

Jet velocity

$$C_1 = 0.96(2 \times 9.81 \times 246)^{1/2}$$

$$= 66.7 \, \text{m/s}$$

Bucket speed

$$U = 0.46 \times 66.7$$

$$= 30.7 \, \text{m/s}$$

Wheel diameter

$$D = \frac{60 \times U}{\pi N}$$

$$= \frac{60 \times 30.7}{\pi \times 650}$$

$$= 0.901 \, \text{m}$$

(c) Overall efficiency

$$\eta_o = \frac{\text{Shaft power developed}}{\text{Power available per wheel}}$$

$$\text{Power per wheel} = \frac{18 \times 10^6}{0.87 \times 6}$$

$$= 3.448 \, \text{MW}$$

Also

$$\text{Power available per wheel} = 0.5 \, mC_1^2$$

$$= 0.5 \rho Q (2gH) C_v^2$$

Thus
$$3.448 \times 10^6 = 10^3 \times Q \times 9.81 \times 246 \times 0.96^2$$

and flow rate
$$Q = 1.55 \, \text{m}^3/\text{s}$$

Discharge coefficient
$$C_d = \frac{\text{Actual nozzle discharge}}{\text{Theoretical nozzle discharge}}$$

$$\frac{\pi d^2 (2gH)^{1/2}}{4} = \frac{Q}{C_d} \qquad \text{where } d \text{ is the nozzle diameter}$$

$$d = \left(\frac{1.55 \times 4}{0.94 \times \pi}\right)^{1/2} \times \left[\frac{1}{(2 \times 9.81 \times 246)^{0.25}}\right]$$

Nozzle diameter $d = 0.174 \, \text{m}$

(d) Total discharge for six machines $= 1.55 \times 6$
$$= 9.3 \, \text{m}^3/\text{s}$$

Total discharge or flow per pipe $= 9.3/3$
$$= 3.1 \, \text{m}^3/\text{s}$$

The frictional head loss in the pipe is given by
$$h_f = \frac{4flv^2}{2gd}$$

where v is the flow velocity and d the pipe diameter. Whence
$$d = \frac{4 \times 0.006 \times 2500v^2}{2 \times 9.81 \times 34}$$

$$= 0.09v^2 \, \text{m}$$

But
$$Q = vA = v\frac{\pi d^2}{4}$$

and substituting for v

$$d = 0.09\left(\frac{3.1 \times 4}{\pi d^2}\right)^2$$

and
$$d^5 = 1.4$$
$$d = 1.07 \, \text{m}$$

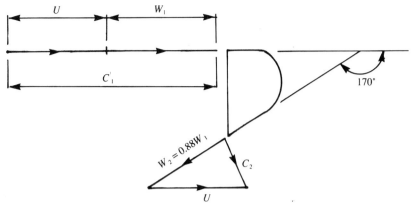

Figure 3.19 Inlet velocity triangle

Exercise 3.3 Figure 3.19 illustrates the system with the velocity triangles. From Eq. (3.4)

$$\text{Hydraulic efficiency} = \frac{\text{Power output}}{\text{Energy available in jet}}$$

$$= \frac{W}{0.5mC_1^2}$$

At entry to nozzle

$$H = 600 - 48$$
$$= 552\,\text{m}$$

Nozzle velocity coefficient

$$C_v = \frac{C_1}{\text{Theoretical velocity}}$$

$$= \frac{C_1}{(2gH)^{1/2}}$$

Thus

$$C_1 = 0.98(2 \times 9.81 \times 552)^{1/2}$$
$$= 102\,\text{m/s}$$

Now

$$W/m = U_1 C_{x1} - U_2 C_{x2}$$
$$= U\{(U + W_1) - [U - W_2 \cos(180° - \alpha)]\}$$
$$= U[(C_1 - U)(1 - k \cos \alpha)]$$

where $W_2 = kW_1$. Substituting for the symbols

$$W/m = 0.47C_1(C_1 - 0.47C_1)(1 - 0.88 \cos 170°)$$
$$W = 0.465mC_1^2$$

$$\text{Theoretical hydraulic efficiency} = 0.465/0.5$$
$$= 0.93$$

$$\text{Actual hydraulic efficiency} = 0.9 \times 0.93$$
$$= 0.837$$

Wheel bucket speed

$$U = 0.47 \times 102 \, \text{m/s}$$

Wheel rotational speed

$$N = \frac{0.47 \times 102 \times 60}{0.45 \times 2\pi}$$

$$= 1017 \, \text{rpm}$$

$$\text{Actual hydraulic efficiency} = \frac{\text{Actual power developed}}{\text{Energy available in jet}}$$

$$0.837 = \frac{1250 \times 10^3}{0.5mC_1^2}$$

Substituting for C_1 and solving for the mass flow rate

$$m = \frac{1250 \times 10^3}{0.837 \times 0.5 \times 102^2}$$

$$= 287 \, \text{kg/s}$$

Hence for one nozzle

$$m = 143.5 \, \text{kg/s}$$

Also from continuity

$$m = \rho C_1 A \qquad \text{where } A \text{ is the nozzle area}$$

$$= \frac{\rho C_1 \pi d^2}{4} \qquad \text{where } d \text{ is the nozzle diameter}$$

and hence

$$d^2 = \frac{143.5 \times 4}{\pi \times 102 \times 10^3}$$

$$= 1.792 \times 10^{-3} \, \text{m}^2$$

$$d = 42.3 \, \text{mm}$$

Exercise 3.4 The angular momentum equation may be used.

Torque

$$T = m(r_2 C_{x2} - r_1 C_{x1})$$

But since the flow is radial at outlet, $C_{x2} = 0$ and therefore

$$T = - mr_1 C_{x1}$$
$$= - \rho Q r_1 C_{x1}$$
$$= - 10^3 \times 0.5 \times 0.6 C_{x1}$$
$$= - 300 C_{x1} \, \text{N m}$$

The inlet area A is

$$A = 2\pi r_1 b_1 \qquad \text{where } b_1 \text{ is the inlet runner height}$$
$$= 2\pi \times 0.6 \times 0.03$$
$$= 0.113 \, \text{m}^2$$

Now flow velocity C_{r1} is given by

$$C_{r1} = Q/A$$
$$= \frac{0.5}{0.113}$$
$$= 4.42 \, \text{m/s}$$

From Fig. 3.12

$$C_{x1} = C_{r1} \tan 72°$$
$$= 4.42 \times 3.08$$
$$= 13.6 \, \text{m/s}$$

Substituting for C_{x1} gives

$$T = - 300 \times 13.6$$
$$= - 4080 \, \text{N m}$$

This is the torque exerted on the control volume (i.e. on the fluid). The torque exerted by the fluid is $+4080 \, \text{N m}$ and is the torque exerted on the runner.

Torque exerted by water on runner $= 4080 \, \text{N m}$

Power exerted

$$W = T\omega$$
$$= \frac{4080 \times 2\pi \times 1260}{60}$$
$$= 538 \, \text{kW}$$

Hydraulic efficiency

$$\eta_H = \frac{\text{Power exerted}}{\text{Power available}}$$

$$= \frac{538 \times 10^3}{\rho g Q H}$$

$$= \frac{538 \times 10^3}{10^3 \times 9.81 \times 0.5 \times 124}$$

$$= 0.885$$

$$= 88.5 \text{ per cent}$$

Exercise 3.5 Figure 3.12 shows the velocity diagrams at inlet and outlet. Hydraulic efficiency

$$\eta_H = \frac{\text{Power given to runner}}{\text{Power available}}$$

$$= \frac{m(U_1 C_{x1} - U_2 C_{x2})}{\rho g Q H}$$

But since the flow is radial at outlet, C_{x2} is zero and m equals ρQ. Therefore

$$\eta_H = \frac{U_1 C_{x1}}{gH}$$

$$0.82 = \frac{0.97(2gH)^{1/2} C_{x1}}{gH}$$

whence

$$C_{x1} = 0.423(2gH)^{1/2}$$

Now

$$\tan \alpha_1 = C_{r1}/C_{x1}$$

$$= 0.4/0.423$$

from which inlet guide vane angle

$$\alpha_1 = 43.4°$$

$$\tan \beta_1 = C_{r1}/W_{x1}$$

$$= \frac{C_{r1}}{C_{x1} - U_1}$$

$$= \frac{0.4}{0.423 - 0.97}$$

$$= -0.731$$

from which $\beta_1 = -36.2°$ to give the blade angle β_1 as $(180° - 36.2°) = \underline{143.8°}$
with $U_1 > C_{x1}$.

Runner speed

$$U_1 = \frac{\pi D_1 N}{60}$$

$$D_1 = \frac{60 \times 0.97 \times (2 \times 9.81 \times 5.5)^{1/2}}{\pi \times 230}$$

Runner inlet diameter $= 0.836\,\mathrm{m}$

Overall efficiency

$$\eta_o = \frac{\text{Power output}}{\text{Power available}}$$

or

$$\rho g Q H = \frac{125 \times 10^3}{0.74}$$

whence flow rate

$$Q = \frac{125 \times 10^3}{0.74 \times 10^3 \times 9.81 \times 5.5}$$

$$= 3.13\,\mathrm{m}^3/\mathrm{s}$$

But also

$$Q = \pi D_1 b_1 C_{r1}$$

Therefore

$$b_1 = \frac{3.13}{\pi \times 0.836 \times 0.4(2 \times 9.81 \times 5.5)^{1/2}}$$

Height of runner $= 0.287\,\mathrm{m}$

Exercise 3.6 Use the notation of Fig. 3.12.

(a) Runner tip speed

$$U_1 = \frac{\pi N D}{60}$$

$$= \frac{\pi \times 430 \times 1.4}{60}$$

$$= 31.5\,\mathrm{m/s}$$

Power given to runner

$$W = m(U_1 C_{x1} - U_2 C_{x2}) \qquad \text{(from Eq. (1.24))}$$

But $C_{x2} = 0$ since there is zero whirl at outlet. Hence

$$C_{x1} = \frac{12\,250 \times 10^3 \times 60}{10^3 \times 12 \times \pi \times 1.4 \times 430}$$

$$= 32.4\,\text{m/s}$$

Guide vane angle

$$\alpha_1 = \tan^{-1}\left(\frac{C_{r1}}{C_{x1}}\right)$$

$$= \tan^{-1}\left(\frac{9.5}{32.4}\right)$$

$$= \underline{16.3^\circ}$$

Inlet velocity

$$C_1 = (C_{r1}^2 + C_{x1}^2)^{1/2}$$
$$= (9.5^2 + 32.4^2)^{1/2}$$
$$= \underline{33.8\,\text{m/s}}$$

(b)

$$\tan \beta_1 = \frac{C_{r1}}{C_{x1} - U_1}$$

$$= \frac{9.5}{32.4 - 31.5}$$

$$= 10.55$$

whence

Runner blade entry angle $\underline{\beta_1 = 84.6^\circ}$

(c) Total head across runner
    = Energy (head) transferred to runner + Head lost in runner

At inlet

$$H_1 = \frac{p_1}{\rho g} + \frac{C_1^2}{2g} + Z_1$$

At outlet

$$H_2 = \frac{p_2}{\rho g} + \frac{C_2^2}{2g} + Z_2$$

Now for zero whirl at outlet

$$\frac{W}{mg} = \frac{U_1 C_{x1}}{g}$$

Hence loss of head in the runner

$$= \left(\frac{p_1 - p_2}{\rho g} \right) + \left(\frac{C_1^2 - C_2^2}{2g} \right) + (Z_1 - Z_2) - \frac{U_1 C_{x1}}{g}$$

But

$$\left(\frac{p_1 - p_2}{\rho g} \right) + (Z_1 - Z_2) = 62 \, \text{m}$$

$$\text{Head loss in runner} = 62 + \left(\frac{33.8^2 - 7^2}{2 \times 9.81} \right) - \left(\frac{31.5 \times 32.4}{9.81} \right)$$

$$= 13.69 \, \text{m}$$

Exercise 3.7 (a) From Eq. (3.21) the hydraulic efficiency is given by

$$\eta_H = \frac{\text{Power transferred to runner}}{\text{Power available}}$$

$$\text{Power transferred to runner} = \rho g Q H \eta_H$$
$$= W \eta_H$$

But from Euler's turbine equation (Eq. (1.24))

$$E = \frac{W}{mg} = \frac{U_1 C_{x1} - U_2 C_{x2}}{g}$$

Therefore

$$\eta_H g H = U_1 C_{x1} - U_2 C_{x2}$$

and $C_{x2} = 0$. Hence

$$C_{x1} = \frac{0.9 \times 9.81 \times 62}{U_1}$$

Now

$$U_1 = \frac{\pi N D}{60}$$

$$= \frac{\pi \times 375 \times 1.5}{60}$$

$$= 29.45 \, \text{m/s}$$

Therefore

$$C_{x1} = \frac{0.9 \times 9.81 \times 62}{29.45}$$

$$= 18.58 \, \text{m/s}$$

The velocity triangle is therefore as shown in Fig. 3.20 with $U_1 > C_{x1}$.

$$W_{x1} = U_1 - C_{x1}$$
$$= 29.45 - 18.58$$
$$= 10.86 \, \text{m/s}$$
$$\tan(\beta_1 - 90°) = W_{x1}/C_{r1}$$
$$= \frac{10.86}{9}$$
$$= 1.21 \, \text{m/s}$$
$$\beta_1 - 90° = 50.4°$$

Entry angle

$$\underline{\beta_1 = 140.4°}$$

(b) (i) For the volute casing and guide vane loss, apply the steady flow energy equation between points 0 and 1:

$$\frac{p_0}{\rho g} + \frac{V_0^2}{2g} + Z_0 = \frac{p_1}{\rho g} + \frac{V_1^2}{2g} + Z_1 + h_{\text{loss},1}$$

Now $V_1 = C_1$ and

$$C_1^2 = C_{x1}^2 + C_{r1}^2$$
$$= 18.58^2 + 9^2$$
$$= 426.2 \, \text{m}^2/\text{s}^2$$

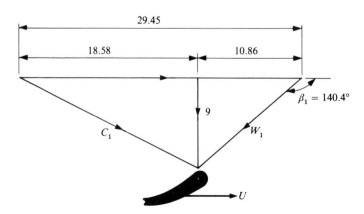

Figure 3.20 Velocity triangle at inlet

Substituting values

$$62 = \left(35 + \frac{426.2}{2 \times 9.81} + 2\right) + h_{\text{loss},1}$$

whence

$$h_{\text{loss},1} = 62 - 58.72$$

$$\underline{\text{Head lost in casing} = 3.27\,\text{m}}$$

(ii) For the loss in the runner, apply the steady flow energy equation between points 1 and 2:

$$\frac{p_1}{\rho g} + \frac{V_1^2}{2g} + Z_1 = \frac{p_2}{\rho g} + \frac{V_2^2}{2g} + Z_2 + h_{\text{loss},2} + \text{Work (head) given to runner}$$

Now

$$\frac{W}{mg} = \frac{U_1 C_{x1}}{g}$$

$$= \frac{29.45 \times 18.58}{9.81}$$

$$= 55.8\,\text{m}$$

Therefore

$$h_{\text{loss},2} = \left(35 + \frac{426.2}{2 \times 9.81} + 2\right) - \left(-2.2 + \frac{7^2}{2 \times 9.81} + 1.7\right) - 55.8$$

$$= 58.72 - 2.0 - 55.8$$

$$\underline{\text{Runner head loss} = 0.92\,\text{m of water}}$$

(iii) Applying the steady flow energy equation between points 2 and 3:

$$\frac{p_2}{\rho g} + \frac{V_2^2}{2g} + Z_2 = \frac{p_3}{\rho g} + \frac{V_3^2}{2g} + Z_3 + h_{\text{loss},3}$$

But p_3 is atmospheric pressure (zero gauge) and Z_3 is the datum level, whence

$$h_{\text{loss},3} = 2.0 - \left(0 + \frac{2.0^2}{2 \times 9.81} + 0\right)$$

$$= 2.0 - 0.204$$

$$\underline{\text{Head loss in draft tube} = 1.8\,\text{m}}$$

(c) Dimensionless power specific speed from Eq. (1.20) is

$$N_{\text{sp}} = \frac{NP^{1/2}}{\rho^{1/2}(gH)^{5/4}}$$

$$P^{1/2} = \frac{0.14 \times (10^3)^{1/2} \times (9.81 \times 62)^{5/4} \times 60}{375}$$

$$= 2140$$

$$P = 4578\,\text{kW} \qquad \text{(this is the power delivered to the runner)}$$

Thus

$$0.9 = \frac{4578 \times 10^3}{\rho g Q H}$$

Flow rate

$$Q = \frac{4578 \times 10^3}{10^3 \times 9.81 \times 62}$$

$$= 7.53\,\text{m}^3/\text{s}$$

$$\text{Flow area} = \frac{\text{Flow rate}}{\text{Flow velocity}}$$

$$= \frac{Q}{C_2} \qquad \text{since } C_2 = C_{r2}$$

and at exit from the runner the flow area may be written in terms of the runner exit diameter and runner height b_2:

$$\pi d_2 b_2 = Q/C_2$$

where d_2 is the draft tube entry diameter. Now the runner height at entry b_1 is given by

$$b_1 = \frac{Q}{\pi d_1 C_{r1}}$$

$$= \frac{7.53}{\pi \times 1.5 \times 9}$$

$$= 0.178\,\text{m}$$

Also

$$b_2 = 2 + \frac{b_1}{2} - 1.7$$

$$= 2 + 0.089 - 1.7$$

$$= 0.389\,\text{m}$$

Substituting for b_2 we get

$$d_2 = \frac{7.53}{\pi \times 0.389 \times 7}$$

$$\underline{\text{Draft tube diameter} = 0.88\,\text{m}}$$

Exercise 3.8 Mean diameter

$$d_m = \frac{D + d}{2}$$

$$= \frac{4.75 + 2}{2}$$

$$= 3.375\,\text{m}$$

Overall efficiency

$$\eta_o = \frac{\text{Power developed}}{\text{Power available}}$$

$$\text{Available power} = \frac{23 \times 10^6}{0.85}$$

$$= 27\,\text{MW}$$

Also

$$\text{Available power} = \rho g Q H$$

$$27 \times 10^6 = 10^3 \times 9.81 \times 23 Q$$

Whence flow rate

$$Q = 119.7\,\text{m}^3/\text{s}$$

Rotor speed at mean diameter

$$U_m = \frac{\pi N d_m}{60}$$

$$= \frac{\pi \times 150 \times 3.375}{60}$$

$$= 26.5\,\text{m/s}$$

$$\text{Power given to runner} = \text{Power available} \times \eta_H$$

$$= 27 \times 10^6 \times 0.93$$

$$= 25.11\,\text{MW}$$

But theoretical power given to runner is from Eq. (1.24)

$$W = \rho Q U_m C_{x1} \qquad (C_{x2} = 0)$$

$$25.11 \times 10^6 = 10^3 \times 119.7 \times 26.5 C_{x1}$$

$$C_{x1} = \frac{25.11 \times 10^6}{10^3 \times 119.9 \times 26.5}$$

$$= 7.9\,\text{m/s}$$

Axial velocity

$$C_a = \frac{Q \times 4}{\pi(D^2 - d^2)}$$

$$= \frac{119.7 \times 4}{\pi(4.75^2 - 2^2)}$$

$$= 8.21 \text{ m/s}$$

From the inlet velocity triangle

$$\tan(180° - \beta_1) = \frac{C_a}{U_m - C_{x1}}$$

$$= \frac{8.21}{26.5 - 7.9}$$

Inlet blade angle $\beta_1 = 156.2°$

At outlet

$$\tan\beta_2 = C_a/W_{x2}$$

But W_{x2} equals U_m since C_{x2} is zero. Hence

$$\tan\beta_2 = \frac{8.21}{26.5}$$

Outlet blade angle $\beta_2 = 17.2°$

Exercise 3.9 Using Eq. (1.20) for power specific speed,

$$N_{sp} = \frac{NP^{1/2}}{\rho^{1/2}(gH)^{5/4}} \qquad \text{where } N \text{ is in rad/s}$$

$$N = 3 \times \left(\frac{10^3}{16\,000 \times 10^3}\right)^{1/2} (9.81 \times 20)^{5/4}$$

$$= 17.41 \text{ rad/s}$$

$$= \frac{17.41 \times 60}{2\pi} \text{ rpm}$$

$$= 166.3 \text{ rpm}$$

Overall efficiency

$$\eta_o = \frac{\text{Power developed}}{\text{Power available}}$$

Therefore

$$\text{Available power} = \frac{16\,000 \times 10^3}{0.8}$$

$$= 20\,000\,\text{kW}$$

Also

$$\text{Available power} = \rho g Q H$$

Therefore

$$Q = \frac{20\,000 \times 10^3}{10^3 \times 9.81 \times 20}$$

$$= 101.9\,\text{m}^3/\text{s}$$

$$\text{Power given to impeller} = \text{Power available} \times \eta_{\text{H}}$$

$$= 20\,000 \times 10^3 \times 0.9$$

$$= 18\,000 \times 10^3\,\text{kW}$$

Now from Eq. (1.24) the power transferred per unit mass flow is

$$W/m = U_1 C_{x1} - U_2 C_{x2}$$

and putting $C_{x2} = 0$ since at exit flow is purely axial, and writing U_1 in terms of the radius at the blade tip, i.e. at 2.1 m,

$$18\,000 \times 10^3 = 17.41 \times 2.1 \times 10^3 \times 101.9 \times C_{x1}$$

$$C_{x1} = 4.8\,\text{m/s}$$

Mean flow velocity

$$C_a = \frac{Q \times 4}{\pi(D^2 - d^2)} \qquad \text{where } D \text{ and } d \text{ are tip and hub diameters}$$

$$= \frac{101.9 \times 4}{\pi(4.2^2 - 2^2)}$$

$$= 9.51\,\text{m/s}$$

From Fig. 3.16

$$W_{x1} = U - C_{x1} \qquad \text{at the blade tip}$$

$$= (17.41 \times 2.1) - 4.8$$

$$= 31.8\,\text{m/s}$$

and

$$\tan(180° - \beta_1) = C_a/W_{x1}$$

$$= \frac{9.51}{31.8}$$

At $r = 2.1$ m

$$\text{Inlet angle } \beta_1 = 163.4°$$

At $r = 2.1$ m

$$W_{x2} = U \quad \text{(since flow is axial at exit)}$$
$$= 17.41 \times 2.1$$
$$= 36.6 \text{ m/s}$$

and

$$\tan \beta_2 = C_a / W_{x2}$$
$$= \frac{9.51}{36.6}$$

At $r = 2.1$ m

$$\text{Outlet angle } \beta_2 = 14.5°$$

At the hub, $r = 1$ m

$$W_{x1} = 17.41 - (4.8 \times 2.1)$$
$$= 7.33 \text{ m/s}$$

Whence

$$(180° - \beta_1) = \tan^{-1}\left(\frac{C_a}{W_{x1}}\right)$$
$$= \tan^{-1}\left(\frac{9.51}{7.33}\right)$$

At $r = 1$ m

$$\text{Inlet angle } \beta_1 = 127.6°$$

At $r = 1$ m

$$W_{x2} = U$$
$$= 17.41 \text{ m/s}$$

Whence

$$\beta_2 = \tan^{-1}\left(\frac{9.51}{17.41}\right)$$

At $r = 1$ m

$$\text{Outlet angle } \beta_2 = 28.6°$$

The point to note in this problem is that we have assumed the energy transfer across the annulus to remain constant and that the axial flow velocity remains constant.

Exercise 3.10 (a) The flow rate is calculated at the mean diameter and the velocity triangles of Fig. 3.16 are used.

Mean diameter

$$d_m = \frac{D+d}{2}$$

$$= \frac{2+0.8}{2}$$

$$= 1.4\,\text{m}$$

Radial velocity

$$U_m = \frac{\pi N d_m}{60}$$

$$= \frac{\pi \times 250 \times 1.4}{60}$$

$$= 18.33\,\text{m/s}$$

From the inlet velocity triangle

$$U_m = C_a[\cot \alpha_1 + \cot (180° - \beta_1)]$$

Axial velocity

$$C_a = \frac{18.33}{\cot 42° + \cot 32°}$$

$$= \frac{18.33}{1.11 + 1.6}$$

$$= 6.76\,\text{m/s}$$

Flow rate

$$Q = \frac{C_a \pi (D^2 - d^2)}{4} \qquad \text{since } C_a \text{ is constant}$$

$$= \frac{6.76 \times \pi (2^2 - 0.8^2)}{4}$$

$$= 17.84\,\text{m}^3/\text{s}$$

(b) From the outlet velocity triangle

$$\tan \beta_2 = C_a/U_m$$

$$= \frac{6.76}{18.33}$$

Outlet blade angle

$$\underline{\beta_2 = 20.24°}$$

(c) The theoretical power is calculated by determining the energy transfer through an annulus of thickness dr and integrating between the hub and tip radii for the total theoretical power.

From Eq. (1.24)

$$W = m(UC_{x1} - UC_{x2})$$

But C_{x2} is zero and C_{x1} is constant and writing m and U as functions of the radius we get for an annulus of thickness dr:

$$dW = \rho C_a (2\pi r \, dr)\left(\frac{2\pi Nr}{60}\right)C_{x1}$$

Integrating

$$W = \frac{\rho C_a \pi^2 N C_{x1}}{15}\int_{0.4}^{1} r^2 \, dr$$

$$= \frac{10^3 \times 6.76 \times \pi^2 \times 250 \times 6.76 \times \cot 42°}{15}\left[\frac{r^3}{3}\right]_{0.4}^{1}$$

$$= 8348(0.333 - 0.0213)10^3$$

Theoretical power = 2602 kW

If the power is calculated at the mean diameter

$$W = mU_m C_{x1}$$
$$= 10^3 \times 17.84 \times 18.33 \times 6.76 \times \cot 42°$$
$$= 2455 \text{ kW}$$

The difference is 6 per cent.

Exercise 3.11 (a) Dimensionless power specific speed is given by Eq. (1.20)

$$N_{sp} = \frac{NP^{1/2}}{\rho^{1/2}(gH)^{5/4}}$$

$$= \frac{300}{60}\left(\frac{2 \times 10^6}{10^3}\right)^{1/2}\frac{1}{(9.81 \times 50)^{5/4}}$$

$$= 0.0969 \text{ rev}$$

From Eq. (3.25)

$$\sigma = \left[\left(\frac{p_{atm} - p_{vap}}{\rho g}\right) - Z_2\right]\Bigg/ H$$

$$= \left[\left(10.3 - \frac{4 \times 10^3}{10^3 \times 9.81}\right) - 6\right]\Bigg/ 50$$

$$= 0.0778$$

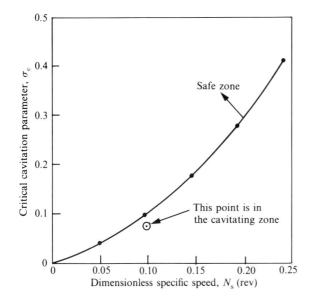

Figure 3.21 Critical cavitation parameter

The characteristics are plotted in Fig. 3.21 and it is seen that cavitation will occur.

(b) The limiting height above the tailrace for no cavitation is when $\sigma_c = 0.1$. Solving for Z_2

$$50 \times 0.1 = 10.3 - \frac{4 \times 10^3}{9.81 \times 10^3} - Z_2$$

when limiting height

$$Z_2 = 10.3 - 0.408 - 5$$
$$= 4.89 \, \text{m}$$

(c) Using the notation of Fig. 3.11 between sections 2 and 3:

$$\frac{p_2}{\rho g} + \frac{V_2^2}{2g} + Z_2 = \frac{p_3}{\rho g} + \frac{V_3^2}{2g} + Z_3 + \text{Losses}$$

But

$$p_3 = p_{\text{atm}}$$
$$V_3 = 0 \quad \text{(negligible)}$$
$$Z_3 = 0 \quad \text{(datum)}$$
$$p_2 = p_{\text{vap}}$$

Therefore the head lost in the draft tube is

$$h_d = \left(\frac{p_{vap} - p_{atm}}{\rho g} \right) + Z_2 + \frac{V_2^2}{2g}$$

At the limit of cavitation $Z_2 = 4.89$ m and substituting into the above equation

$$h_d = \frac{4 \times 10^3}{10^3 \times 9.81} - 10.3 + 4.89 + \frac{10.4^2}{2 \times 9.81}$$

$$= 0.408 - 10.3 + 4.89 + 5.51$$

Head loss $h_d = 0.511$ m

FOUR

CENTRIFUGAL COMPRESSORS AND FANS

4.1 INTRODUCTION

During the Second World War, great progress was made in the development of gas turbines using the centrifugal compressor. This followed from the need for gas turbines to be supplied with large amounts of high-pressure air, and the centrifugal compressor became a natural choice as it had previously been researched for use in small high-speed internal combustion engines. Although the centrifugal compressor has been superseded by the axial flow compressor in jet aircraft engines, it is useful where a short overall engine length is required and where it is likely that deposits will be formed in the air passages, since, because of the relatively short passage length, loss of performance due to build-up of deposits will not be as great as in the axial compressor. The centrifugal compressor is mainly found in turbochargers, where it is placed on the same shaft as an inward flow radial gas turbine, which is driven by engine exhaust gases (Fig. 4.1). Pressure ratios of 4:1 are typical in a single stage, and ratios of 7:1 are possible if exotic materials are used for impeller manufacture. The best efficiencies are 3-4 per cent below those obtainable from an axial flow compressor designed for the same duty. However, at very low mass flow rates the axial flow compressor efficiency drops off rapidly. It is also difficult to hold the tolerances required for small axial flow blading, and manufacture of the axial compressor blades becomes more expensive. If the density ratio across the compressor is less than about 1.05, the term 'fan' is used to describe the machine. In that case the fluid is treated as being incompressible; otherwise compressible flow equations must be used. The term 'blower' is often used in place of 'fan'.

4.1.1 Elements of the Centrifugal Compressor

The elements of a centrifugal compressor are similar to those of a hydraulic pump, with some important differences, and these are illustrated in Fig. 4.1.

The gas enters the compressor at the eye in an axial direction with absolute velocity C_1, and moves into the inducer section, which can be separate from or be a part of the blades. The inducer section transfers the gas onto the blades, and enables it to move smoothly into the radial direction. Energy is imparted to the gas by the rotating blades, thereby increasing its static pressure as it moves from radius r_1 to r_2 (Fig. 4.2), and the gas moves off the blades with absolute velocity C_2. It should be noted that the blades are radial, and since it is conventional to measure blade angles from the radial direction in centrifugal compressors, the blade angle β_2 is zero, while the relative velocity vector W_2 is at angle β'_2 because of slip. Ideally, the component C_{x2} equals U_2, but it is reduced due to slip. The relative velocity vector W_1 at the inlet is obtained by subtracting U_1 from C_1. Pre-whirl may be given to the gas at inlet, but this will be discussed in a later section. The Stanitz

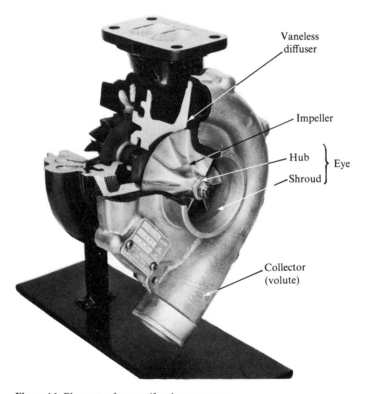

Figure 4.1 Elements of a centrifugal compressor

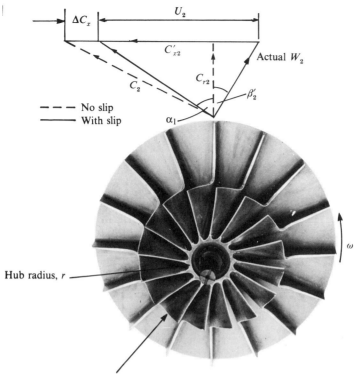

Hub radius, r

Eye tip or shroud radius, R

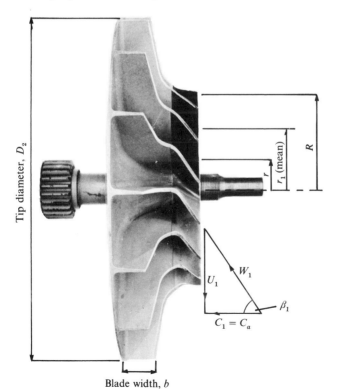

Figure 4.2 Velocity diagrams for a centrifugal compressor

slip factor (Eq. (2.9)) is best applied for radial vanes and with $\beta_2 = 0$,

$$\sigma_s = 1 - 0.63\pi/Z$$

By Euler's pump equation (Eq. (1.25)) without slip,

$$E = U_2 C_{x2}/g = U_2^2/g \tag{4.1}$$

and with slip,

$$E = \sigma_s U_2^2/g \tag{4.2}$$

Although Eq. (4.1) has been modified by the slip factor to give Eq. (4.2), $\sigma_s U_2^2/g$ is still the theoretical work done on the air, since slip will be present even if the fluid is frictionless. In a real fluid, some of the power supplied by the impeller is used in overcoming losses that have a braking effect on the air conveyed by the vanes, and these include windage, disc friction and casing friction. The total power per unit weight of flow is therefore modified by a power input factor ψ, which typically takes values between 1.035 and 1.040.

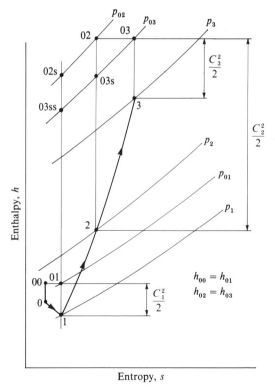

Figure 4.3 Mollier chart for a centrifugal compressor

Thus E becomes

$$E = \psi \sigma_s U_2^2 / g \tag{4.3}$$

Upon leaving the impeller the gas enters a vaneless space where it moves in a spiral path before entering the diffuser, in which the static pressure is further increased. The clearance between the impeller blades and inner walls of the casing must be kept as small as possible to reduce leakage and in some cases the blades themselves are shrouded. Since we are dealing with a gas and since the rise in temperature and pressure causes the density to change, it is convenient to examine the performance of the machine in terms of the thermodynamic properties of the gas, and this is done through the Mollier chart of Fig. 4.3.

4.1.2 Inlet Casing

The energy equation along a streamline may be written as

$$\text{Total enthalpy } h_0 = h + C^2/2 = \text{Constant}$$

Therefore, for the fluid that is being drawn from the atmosphere into the inducer section, the total enthalpy is

$$h_{00} = h_0 + C_0^2/2$$

Total enthalpy at section 1 is

$$h_{01} = h_1 + C_1^2/2$$

and since no shaft work has been done and assuming adiabatic steady flow,

$$h_{00} = h_{01}$$

Thus

$$h_0 + C_0^2/2 = h_1 + C_1^2/2 \tag{4.4}$$

4.1.3 Impeller

From section 1 to 2 the fluid moves through the impeller where work is done on it to increase its static pressure from p_1 to p_2. Writing the work done per unit mass on the fluid in terms of enthalpy we get

$$W/m = h_{02} - h_{01}$$
$$= U_2 C_{x2} - U_1 C_{x1} \tag{4.5}$$

or after substituting for h_0,

$$I = h_1 + C_1^2/2 - U_1 C_{x1} = h_2 + C_2^2/2 - U_2 C_{x2}$$

where I is a constant throughout the impeller. In general

$$
\begin{aligned}
I &= h + C^2/2 - UC_x \\
&= h + (C_r^2 + C_x^2)/2 - UC_x \\
&= h + (W^2 - W_x^2 + C_x^2)/2 - UC_x \\
&= h + [W^2 - (U - C_x)^2 + C_x^2]/2 - UC_x \\
&= h + W^2/2 - U^2/2 - C_x^2/2 + UC_x - UC_x + C_x^2/2 \\
&= h + W^2/2 - U^2/2 \\
&= h_{0,\text{rel}} - U^2/2
\end{aligned}
\tag{4.6}
$$

where $h_{0,\text{rel}}$ is the total enthalpy based on the relative velocity of the fluid. Thus

$$
h_2 - h_1 = (U_2^2 - U_1^2)/2 + (W_1^2 - W_2^2)/2
\tag{4.7}
$$

since $I_1 = I_2$. In Eq. (4.7) the main contribution to the static enthalpy rise is from the term $(U_2^2 - U_1^2)/2$.

In preliminary design calculations it is usual to assume $C_{x1} = 0$, although this is not always the case, whence from Eq. (4.3) the work done on the fluid per unit mass becomes

$$
h_{02} - h_{01} = \psi \sigma_s U_2^2 \ (\text{J/kg})
$$

After writing $C_p T_0$ in place of h_0, we get that the work input is given by

$$
T_{02} - T_{01} = \psi \sigma_s U_2^2 / C_p
\tag{4.8}
$$

where C_p is the mean specific heat over this temperature range. Also, since no work is done in the diffuser, $h_{02} = h_{03}$, and Eq. (4.8) becomes

$$
T_{03} - T_{01} = \psi \sigma_s U_2^2 / C_p
\tag{4.9}
$$

With reference to Figs 1.4 and 4.3, a compressor overall total-to-total isentropic efficiency η_c may be defined as

$$
\eta_c = \frac{\text{Total isentropic enthalpy rise between inlet and outlet}}{\text{Actual enthalpy rise between same total pressure limits}}
$$

$$
= (h_{03ss} - h_{01})/(h_{03} - h_{01})
\tag{4.10}
$$

where the subscript 'ss' represents the end state on the total pressure line p_{03} when the process is isentropic. Thus

$$
\begin{aligned}
\eta_c &= (T_{03ss} - T_{01})/(T_{03} - T_{01}) \\
&= T_{01}(T_{03ss}/T_{01} - 1)/(T_{03} - T_{01})
\end{aligned}
$$

But

$$
\begin{aligned}
p_{03}/p_{01} &= (T_{03ss}/T_{01})^{\gamma/(\gamma-1)} \\
&= [1 + \eta_c(T_{03} - T_{01})/T_{01}]^{\gamma/(\gamma-1)} \\
&= (1 + \eta_c \psi \sigma_s U_2^2 / C_p T_{01})^{\gamma/(\gamma-1)}
\end{aligned}
\tag{4.11}
$$

The slip factor should be as high as possible since it limits the energy transfer to the fluid even under isentropic conditions and it is seen from the velocity diagrams that C_{x2} approaches U_2 as the slip factor is increased. The slip factor may be increased by increasing the number of vanes but this increases the solidity at the impeller eye, resulting in a decrease in the flow area at inlet. For the same mass flow rate, the flow velocity C_a at inlet must therefore be increased and this increases the loss due to friction. A compromise is usually reached, slip factors of about 0.9 being typical for a compressor with 19–21 vanes.

While it may seem that a high value of power input factor ψ is desirable, it is found that the rate of decrease of isentropic efficiency with increase in ψ negates any apparent advantage, so the ideal should be to have a power input factor of unity.

The pressure ratio increases with the impeller tip speed but material strength considerations preclude this being increased indefinitely. Centrifugal stresses are proportional to the square of the tip speed and, for a light alloy impeller, tip speeds are limited to about 460 m/s. This gives a pressure ratio of about 4:1. Pressure ratios of 7:1 are possible if materials such as titanium are used.

Equation (4.11) can be written in terms of fluid properties and flow angles

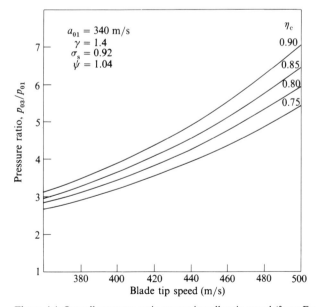

Figure 4.4 Overall pressure ratio versus impeller tip speed (from Eq. (4.12))

as follows: since

$$a_{01}^2 = \gamma R T_{01} \quad \text{and} \quad C_p = \gamma R/(\gamma - 1)$$

then

$$p_{03}/p_{01} = [1 + \eta_c \psi \sigma_s(\gamma - 1)U_2^2/a_{01}^2]^{\gamma/(\gamma - 1)} \quad (4.12)$$

The change of pressure ratio with blade tip speed is shown in Fig. 4.4 for various isentropic efficiencies.

4.1.4 Diffuser

The stagnation temperature of the gas at outlet from the diffuser should have as small a kinetic energy term as possible, as this eases the problem of combustion chamber design. Typical compressor outlet velocities are of the order of 90 m/s. The diffusion process is carried out in a diffuser as described in Secs 2.5.2 and 2.5.3, some diffusion also taking place in the vaneless space between the impeller tip and diffuser vanes. The flow theory described in those sections is applicable here. The maximum included angle of the vaned diffuser passage is about $11°$, any increase in this angle leading to a loss of efficiency through boundary-layer separation on the passage walls. It should also be remembered that any change from the design mass flow rate and pressure ratio will change the smooth flow direction into the diffuser passage and will therefore also result in a loss of efficiency. This may be rectified by utilizing variable-angle diffuser vanes.

For adiabatic deceleration of the fluid from absolute velocity C_2 to C_3 with a corresponding increase of static pressure from p_2 to p_3,

$$h_{02} = h_{03}$$

or

$$h_2 + C_2^2/2 = h_3 + C_3^2/2 \quad (4.13)$$

4.2 INLET VELOCITY LIMITATIONS

Mach-number considerations at the eye of a centrifugal compressor make the relative velocity W_1 a very sensitive value as far as compressor performance is concerned. Should the Mach number at entry to the impeller be greater than unity, then shock waves will form, with all their attendant losses. Assume that we have a uniform absolute velocity C_1 with zero whirl ($C_{x1} = 0$) at entry to a centrifugal compressor. Two cases may be examined for the same mass flow rate, both cases being extremes.

1. If the eye tip diameter is large, then from continuity considerations the axial velocity C_1 is low and the blade speed is high, resulting in the velocity diagram of Fig. 4.5a.

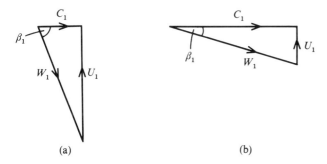

Figure 4.5 Velocity triangles for (a) large and (b) small inlet areas

2. If the eye tip diameter is small, the blade speed is small but the axial velocity C_1 is large, and the velocity diagram of Fig. 4.5b may be drawn.

For both of these extreme cases, the relative velocity vector W_1 is high but it must reach a minimum value when moving from one extreme to another. If this relative velocity can be determined, Mach-number effects can be avoided by proper design.

Flow into the eye takes place through the annulus formed by the shroud radius R and the hub radius r. For uniform axial flow into the eye

$$m = \rho_1 A_1 C_1 \tag{4.14}$$

but from the inlet velocity triangle (Fig. 4.5),

$$C_1 = W_1 \cos \beta_1 \quad \text{and} \quad U_1 = W_1 \sin \beta_1$$

The flow area is

$$A_1 = \pi R^2 (1 - r^2/R^2)$$
$$= \pi R^2 k \tag{4.15}$$

where $k = (1 - r^2/R^2)$. Substitution for A_1 into Eq. (4.14) gives

$$m = \rho_1 \pi R^2 k C_1$$
$$= \rho_1 \pi U_1^2 k C_1/\omega^2$$

where U_1 is the inlet tangential velocity of the impeller at the shroud radius and ω is the angular velocity. Rearranging gives

$$m\omega^2/\rho_1 \pi k = U_1^2 C_1$$
$$= W_1^3 (\sin^2 \beta_1)(\cos \beta_1) \tag{4.16}$$

From isentropic flow relationships the ratio of stagnation to static pressure at inlet may be written as

$$p_{01}/p_1 = [1 + (\gamma - 1)M_1^2/2]^{\gamma/(\gamma - 1)} \tag{4.17}$$

and

$$T_{01}/T_1 = 1 + (\gamma - 1)M_1^2/2 \qquad (4.18)$$

Then

$$p_1/T_1 = (p_{01}/T_{01})[1 + (\gamma - 1)M_1^2/2]^{-\gamma/(\gamma - 1)}[1 + (\gamma - 1)M_1^2/2]$$

$$= (p_{01}/T_{01})[1 + (\gamma - 1)M_1^2/2]^{-1/(\gamma - 1)}$$

Now

$$\rho_1 = p_1/RT_1$$

$$= (p_{01}/RT_{01})[1 + (\gamma - 1)M_1^2/2]^{-1/(\gamma - 1)} \qquad (4.19)$$

Therefore substituting for ρ_1 from Eq. (4.19) into Eq. (4.16),

$$m\omega^2 RT_{01}/\pi k p_{01} = W_1^3(\sin^2 \beta_1)(\cos \beta_1)/[1 + (\gamma - 1)M_1^2/2]^{1/(\gamma - 1)} \quad (4.20)$$

Writing the relative Mach number based on the relative velocity W_1, then

$$m\omega^2 RT_{01}/\pi k p_{01} = M_{1,rel}^3 a_1^3(\sin^2 \beta_1)(\cos \beta_1)/[1 + (\gamma - 1)M_1^2/2]^{1/(\gamma - 1)} \quad (4.21)$$

From Eq. (4.18), $a_{01}/a_1 = [1 + (\gamma - 1)M_1^2/2]^{1/2}$ since $a = (\gamma RT)^{1/2}$ and after substituting for a_1 in Eq. (4.21) and putting $M_1 = M_{1,rel} \cos \beta_1$,

$$(m\omega^2)/[\pi k \gamma p_{01}(\gamma RT_{01})^{1/2}]$$

$$= M_{1,rel}^3(\sin^2 \beta_1)(\cos \beta_1)/[1 + (\gamma - 1)M_{1,rel}^2(\cos^2 \beta_1)/2]^{1/(\gamma - 1) + 3/2} \quad (4.22)$$

It should be remembered that Eq. (4.22) is applied at the shroud radius R and consequently β_1 is also the blade angle at radius R since it is at this radius that the maximum value of relative velocity onto the blade will occur.

Therefore, for a gas of known inlet stagnation conditions (e.g. the

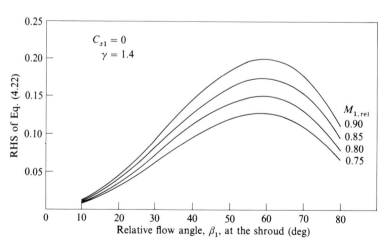

Figure 4.6 Optimization of the mass flow function

atmosphere), the right-hand side of Eq. (4.22) may be plotted and the maximum value determined along with the corresponding blade angle at which the maximum occurs. This maximum value may then be equated to the left-hand side of Eq. (4.22) and the maximum mass flow rate determined. Figure 4.6 shows Eq. (4.22) plotted for air at $p_{01} = 101.3\,\text{kPa}$ and $T_{01} = 288\,\text{K}$ where it is seen that the blade angle is almost constant at $60°$ for maximum mass flow. Therefore, by specifying the relative Mach number $M_{1,\text{rel}}$, the maximum value of mass flow may be calculated. Relative Mach numbers are usually restricted to about 0.8 to ensure there is no shock-wave formation.

4.3 PRE-WHIRL AND INLET GUIDE VANES

Should it not be possible to restrict the Mach number at inlet to an acceptable value as described in Sec. 4.2, it may be achieved by placing guide vanes at the inlet. Figure 4.7 clearly shows that the inlet guide vanes impart a whirl component C_{x1} to the fluid, thus reducing W_1 to an acceptable value. However, the work capacity is reduced since C_{x1} is no longer zero. It is not necessary to impart pre-whirl down to the hub, as, in this region, the fluid is nowhere near sonic conditions due to the lower blade speed. The pre-whirl is therefore gradually reduced to zero by twisting the inlet guide vanes.

4.4 MACH NUMBER IN THE DIFFUSER

The absolute Mach number of the fluid leaving the impeller may well be in excess of 1. However, it has been found that as long as the radial flow velocity C_{r2} is subsonic, then no loss in efficiency is caused by the formation of shock

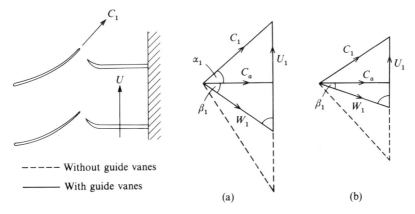

Without guide vanes
With guide vanes

(a)   (b)

Figure 4.7 The effect of inlet guide vanes on the inlet relative velocity: (a) at shroud; (b) at hub

waves. In addition, provided constant angular momentum with vortex motion is maintained in the vaneless space between impeller tip and diffuser, then supersonic diffusion can take place in the vaneless space. This reduces the Mach number at inlet to the diffuser vanes to about 0.8. High Mach numbers at inlet to the diffuser vanes will also cause high pressures at the stagnation points on the diffuser vane tips, which leads to a variation of static pressure around the circumference of the diffuser. This pressure variation is transmitted radially across the vaneless space and can cause cyclic loading of the impeller, which may lead to early fatigue failure.

4.5 CENTRIFUGAL COMPRESSOR CHARACTERISTIC

The characteristics of compressible flow machines are usually described in terms of the groups of variables derived in Eq. (1.14). The characteristics are given as a series of curves of p_{03}/p_{01} plotted against the mass flow parameter $mT_{01}^{1/2}/p_{01}$ for fixed speed intervals of $N/T_{01}^{1/2}$. This might be done by controlling the flow through a valve situated downstream of the compressor outlet flange.

An idealized fixed-speed characteristic is shown in Fig. 4.8. In discussing the shape of this curve, much of what will be said will also apply to the axial flow compressor characteristics of Sec. 5.10.

Consider a centrifugal compressor delivering through a flow control valve situated after the diffuser. If the valve is fully closed, a certain pressure ratio from inlet across to the diffuser outlet will be developed, and this is shown at point 1 in Fig. 4.8. This pressure ratio is due solely to the vanes moving the air about in the impeller. As the flow control valve is opened, and air flow begins, the diffuser contributes to the pressure ratio, and at point 2 the maximum

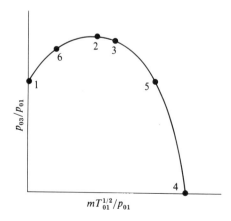

Figure 4.8 The theoretical centrifugal compressor characteristic

pressure ratio is reached, although the compressor efficiency at this pressure ratio will be just below the maximum efficiency. A further increase in mass flow brings conditions to point 3, where the pressure has dropped slightly from the maximum value but the efficiency is now a maximum. This is likely to be the design mass flow rate pressure ratio. A further increase in mass flow sees the slope of the curve increasing until it is almost vertical at point 4 where the pressure rise is zero. Theoretically, point 4 would be reached when all the input power is absorbed in overcoming internal friction. However, the curve just described is not obtainable practically and some of the reasons for this are now discussed.

4.5.1 Surging

Suppose we have a compressor operating at point 3 on the negative slope of the characteristic (Fig. 4.8). A reduction in mass flow due, say, to a momentary blockage will cause an increase in the delivery pressure p_{03}, which will tend to return the mass flow to point 3, and this negative slope represents a region of stable operation. It is self-correcting. If the flow rate drops to a point corresponding to 6 on the positive slope of the characteristics, the delivery pressure p_{03} will continue to decrease, causing a further drop in mass flow and a further drop in p_{03}, and so on until point 1 is reached, where the mass flow is zero. The mass flow may even become negative through the compressor. When the back-pressure p_{03} has reduced itself sufficiently due to the reduced flow rate, the positive flow becomes established once again and the compressor picks up until the restricted mass flow is again reached, when pressure reduction once again takes place. The pressure therefore surges back and forth in an unstable fashion, which, if severe enough, could lead to failure of parts of the compressor. Because of the reduction of mass flow, the axial velocity C_a into the eye is reduced and therefore the relative flow angle onto the blade β_1 is increased. The air flow onto the blade will no longer be tangential. Surging tends to originate in diffuser passages where frictional effects of the fluid next to the vane surfaces retard the flow. Indeed, flow may well be in reverse from one blade passage to the next. The likelihood of surging can be reduced by making the number of diffuser vanes an odd-number multiple of the impeller vanes. In this way a pair of diffuser passages will be supplied with air from an odd number of vanes and pressure fluctuations are more likely to be evened out around the circumference than if exact multiples of diffuser vanes are employed.

4.5.2 Rotating Stall

Rotating stall is a separate phenomenon, which may lead to surging but can exist on its own in a stable operating condition. Figure 4.9 illustrates the air flow directions in a number of blade passages.

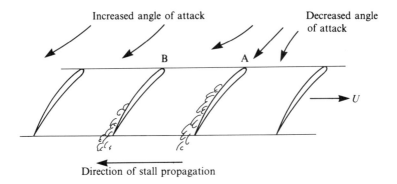

Figure 4.9 Rotating stall propagation

If the air angle of incidence onto blade A is excessive, due perhaps to a partial blockage or uneven flow in the diffuser, the blade may stall, and because of the partial mass flow decrease in the blade passage, the deflected air causes the angle of incidence to the left of blade A to increase while the angle of incidence to the right of blade A will tend to decrease. Thus blade B will be the next to stall while blade A will be unstalled and the process is repeated about the periphery of the disc. Prolonged cyclic loading and unloading of the rotor blades can lead to fatigue failure or even immediate catastrophic failure. The stall propagates in the opposite direction to blade motion at a frequency related to shaft speed. In compressor tests, rotating stall may be audibly recognized as a high-frequency 'screech'.

4.5.3 Choking

If the mass flow is increased to the right of point 3 on the negative slope of the characteristic (Fig. 4.8), a point 5 is reached where no further increase in mass flow is possible no matter how wide open the flow control valve is. This indicates that at some point within the compressor sonic conditions have been reached, causing the limiting maximum mass flow rate to be set as in the case of compressible flow through a converging–diverging nozzle. Indeed, should this condition arise, shock waves may well be formed within certain passages.

Choking may take place at the inlet, within the impeller, or in the diffuser section. It will occur in the inlet if stationary guide vanes are fitted, the maximum mass flow being governed by the following standard equation for isentropic flow at the throat of a converging nozzle:

$$m/A = \{\gamma p_{00}\rho_{00}[2/(\gamma + 1)]^{(\gamma + 1)/(\gamma - 1)}\}^{1/2} \tag{4.23}$$

where stagnation conditions at inlet are known and A is the flow area.

In stationary passages, the velocity that is choked is the absolute velocity. In the rotating impeller it is the relative velocity W that is the choked velocity.

Now

$$h_{01} = h_1 + C_1^2/2 \tag{4.24}$$

$$= h_1 + (W_1^2 - U_1^2)/2 \tag{4.25}$$

If choking occurs when the relative velocity equals the acoustic velocity a_1, Eq. (4.25) becomes

$$T_{01} = T_1 + (\gamma R T_1 - U_1^2)/2C_p$$

and

$$T_1/T_{01} = 2[1 + (U_1^2/2C_p T_{01})]/(\gamma + 1) \tag{4.26}$$

For isentropic flow,

$$\rho_1/\rho_{01} = (T_1/T_{01})^{1/(\gamma - 1)}$$

and therefore from $m = \rho a A$

$$m/A = \rho_{01} a_{01}(T_1/T_{01})^{(\gamma + 1)/2(\gamma - 1)}$$

since $a \propto T^{1/2}$. Substituting from Eq. (4.26) and rearranging gives

$$m/A = \rho_{01} a_{01}[2(1 + U_1^2/2C_p T_{01})/(\gamma + 1)]^{(\gamma + 1)/2(\gamma - 1)}$$
$$= \{\gamma p_{01} \rho_{01}[2(1 + U_1^2/2C_p T_{01})/(\gamma + 1)]^{(\gamma + 1)/(\gamma - 1)}\}^{1/2} \tag{4.27}$$

Equation (4.27) is simply a modified version of Eq. (4.23) and shows that the choking mass flow rate increases with impeller speed.

In the diffuser passages Eq. (4.23) also applies with the subscripts changed to the impeller outlet conditions

$$m/A = \{\gamma p_{02} \rho_{02}[2/(\gamma + 1)]^{(\gamma + 1)/(\gamma - 1)}\}^{1/2} \tag{4.28}$$

The areas in Eqs (4.23), (4.27) and (4.28) refer to the flow areas at the respective locations.

4.5.4 Characteristic Curve

Figure 4.10 shows the overall pressure ratio and efficiency plotted against $m T_{01}^{1/2}/p_{01}$ at fixed speed intervals of $N/T_{01}^{1/2}$. It is usual to transfer constant-efficiency points onto the corresponding constant-speed curves of the pressure ratio characteristics and then join those points together to form constant-efficiency contours.

It is evident that at all speeds the range of mass flow over which the centrifugal compressor will operate before surging or choking occurs is quite wide, although the margin decreases as the speed increases. The onset of surge occurs at increasingly high mass flows as the speed increases, while the locus of the limit of stability is called the surge line. Maximum efficiency is seen to occur well within the surge point, making this type of compressor relatively insensitive to mass flow changes as far as instability is concerned. The limit of maximum flow is usually set by choking in the impeller, while the surge limit of

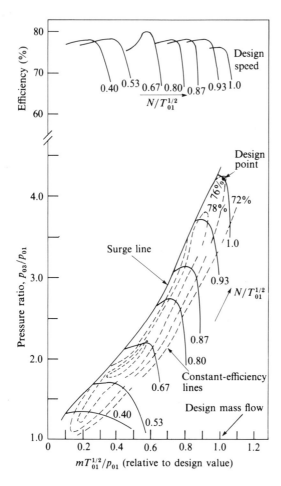

Figure 4.10 Typical centrifugal compressor characteristic

minimum mass flow is set by stalling of the flow onto the diffuser vanes. For a vaneless diffuser the surge line would be further to the left of that shown in Fig. 4.10, thus increasing the mass flow range at the design speed.

EXERCISES

In all the exercises below, the following properties for air may be assumed:
Ratio of specific heats, $\gamma = 1.4$
Specific heat of air at constant pressure, $C_p = 1005\,\text{J/kg K}$
Gas constant, $R = 287\,\text{J/kg K}$
Assume a power input factor of 1 unless otherwise stated.

4.1 A single-sided centrifugal compressor draws air from the atmosphere having ambient conditions of 100 kPa and 15°C. The hub diameter is 0.13 m and the eye tip diameter 0.3 m. If the mass flow into the eye of the compressor is 8 kg/s and the speed is 16 200 rpm, calculate the blade inlet angle at the root and tip of the eye and the Mach number at the tip of the eye. Assume zero whirl at inlet and no losses in the intake duct.

4.2 A motor rated at 580 kW is available to drive a centrifugal compressor of 480 mm outer diameter at a speed of 20 000 rpm. At the impeller outlet the blade angle is 26.5° measured from the radial direction and the flow velocity (radial component) at exit from the impeller is 122 m/s. If a mechanical efficiency of 95 per cent is assumed, what air flow is to be expected? Assume there is no slip. What are the eye tip and hub diameters if a radius ratio of 0.3 is chosen for the impeller eye and if the velocity at inlet is 95 m/s with zero whirl? What will be the overall total-to-total isentropic efficiency if an overall total pressure ratio of 5.5 is required? Assume that the flow to the inlet is incompressible and ambient air conditions are 101.3 kPa and 288 K.

4.3 A centrifugal compressor compresses air at ambient temperature and pressure of 288 K and 101.3 kPa respectively. The impeller runs at a tip speed of 365 m/s, the radial velocity at exit from the impeller is 30 m/s and the slip factor is 0.9. Calculate the Mach number of the flow at the impeller tip. If the impeller total-to-total efficiency is 90 per cent and the flow area from the impeller is 0.093 m^2, calculate the mass flow rate of air. Assume zero whirl at inlet, and radial blades.

4.4 A centrifugal compressor impeller has 17 radial vanes of tip diameter 165 mm. It rotates at 46 000 rpm and the air mass flow rate is 0.6 kg/s with no whirl at inlet. Calculate the theoretical power transferred to the air. At inlet to the impeller, the mean diameter of the eye is 63.5 mm while the annulus height at the eye is 25 mm. The static pressure and temperature at the impeller inlet are 93 kPa and 293 K respectively. Determine

    (a) the blade angle at the mean diameter at impeller inlet,
    (b) the stagnation temperature at impeller exit and
    (c) the stagnation pressure at impeller exit if the total-to-total efficiency of the impeller is 90 per cent.

4.5 The relative flow Mach number at the inlet of a centrifugal compressor is to be no greater than 0.97. The ratio of hub to tip diameter at the eye is 0.38 and the air enters the eye without whirl. Determine

    (a) the optimum speed of rotation for the maximum mass flow rate condition if the mass flow is 4 kg/s and
    (b) the eye tip diameter.
    At inlet $p_{01} = 101.3$ kPa and $T_{01} = 288$ K.

4.6 A compressor operating at a pressure ratio of 3.8 and a speed of 12 000 rpm delivers 8 kg/s of air. The slip factor is assumed to be 0.92, the power factor 1.04, and the overall isentropic efficiency 0.82. Calculate the impeller outlet diameter. Assume zero whirl.

    The Mach number of the air leaving the impeller vanes is to be unity so as to ensure that no shocks occur. If the losses in the impeller and diffuser are the same, what must be the axial depth of the impeller?
    At inlet $p_{01} = 101.3$ kPa and $T_{01} = 288$ K.

4.7 In exercise 4.6, if the vaneless space is 40 mm wide, the diffuser throat radius 0.4 m and there are 10 diffuser channels, determine the diffuser vane inlet angle and the throat width of the diffuser passages.

4.8 A centrifugal compressor has inlet guide vanes fitted at the eye such that free vortex flow is achieved at entry to the blades. At the tip radius of the eye the inlet relative Mach number is not to exceed 0.75 and an impeller total-to-total efficiency of 0.9 is required. The air leaves the tip of the inlet guide vanes with a velocity of 90 m/s, the impeller tip diameter is 0.45 m and the outlet diameter is 0.76 m. The radial component of velocity at exit from the impeller is 50 m/s and the impeller rotates at 12 000 rpm. If a slip factor of 0.9 is assumed, find the guide vane inlet angle at the tip and the static pressure at impeller outlet.

Assume $T_{01} = 288$ K and $p_{01} = 101.3$ kPa.

4.9 A centrifugal compressor has the following design data:

Mass flow	15 kg/s
Speed	16 000 rpm
Inlet total temperature	288 K
Inlet total pressure	101 kPa
Impeller isentropic efficiency	0.9
Width of vaneless space	42 mm
Axial depth of vaneless space	37 mm
Slip factor	0.9
Power input factor	1.04
Impeller outer diameter	0.55 m

(a) Assuming no pre-whirl at the inlet, what are the stagnation conditions at the impeller outlet?

(b) Show that the radial velocity is approximately 100 m/s at the impeller outlet and calculate the Mach number and air angle at the impeller outlet.

(c) Calculate the angle of the diffuser vane leading edges and the Mach number at this radius if the diffusion in the vaneless space is isentropic.

4.10 The stagnation pressure and temperature at the inlet of a centrifugal compressor are 101 kPa and 290 K respectively. The impeller has 19 radial vanes and no inlet guide vanes. The following data apply.

Mass flow rate	2 kg/s
Impeller tip speed	525 m/s
Mechanical efficiency	96 per cent
Absolute air velocity at diffuser exit	95 m/s
Compressor total-to-total efficiency	81 per cent
Absolute velocity at impeller entry	160 m/s
Diffuser efficiency	83 per cent
Axial depth of impeller	6 mm
Power input factor	1.04

(a) Determine the shaft power.

(b) Calculate the total and static pressures at diffuser outlet.

(c) If the reaction ratio $(h_2 - h_1)/(h_3 - h_1)$ is 0.5, find the radial velocity, absolute Mach number and total and static pressures at the impeller exit. Diffuser efficiency $= (h_{3s} - h_2)/(h_3 - h_2)$

(d) Determine the impeller total-to-total efficiency.

(e) If a vaneless diffuser is fitted, what is the ratio of inlet to outlet radius of the diffuser?

(f) Determine the impeller speed.

SOLUTIONS

Exercise 4.1 In Fig. 4.2 the blade angle β_1, measured with respect to the axial velocity C_a, is required to be found at the hub and tip. First the axial velocity must be determined from the continuity equation, but since the inlet density ρ_1 is unknown a trial-and-error procedure must be followed, assuming first a density based on the inlet stagnation conditions. The method is set out below.

Let the hub and tip radii at the eye be denoted by r_h and r_t respectively. Then the flow area of the impeller inlet annulus is

$$A_1 = \pi(r_t^2 - r_h^2)$$
$$= \pi(0.15^2 - 0.065^2)$$
$$= 0.0574 \,\text{m}^2$$

Assume first a value for density ρ_1 based on ambient conditions:

$$\rho_1 = \frac{p_{01}}{RT_{01}}$$
$$= \frac{10^5}{287 \times 288}$$
$$= 1.21 \,\text{kg/m}^3$$

Then from the continuity equation (Eq. (1.21))

$$C_a = \frac{m}{\rho_1 A_1}$$
$$= \frac{8}{1.21 \times 0.0574}$$
$$= 115.18 \,\text{m/s}$$

Since there is no inlet whirl component, $C_a = C_1$, the absolute inlet velocity, and the temperature equivalent of this velocity is

$$\frac{C_1^2}{2C_p} = \frac{115.18^2}{2 \times 1005}$$
$$= 6.6 \,\text{K}$$

Therefore

$$T_1 = T_{01} - \frac{C_1^2}{2C_p}$$
$$= 288 - 6.6$$
$$= 281.4 \,\text{K}$$

Assuming isentropic flow at the inlet

$$\frac{p_1}{p_{01}} = \left(\frac{T_1}{T_{01}}\right)^{\gamma/(\gamma-1)}$$
$$p_1 = 10^5 \left(\frac{281.4}{288}\right)^{1.4/(1.4-1)}$$
$$= 92.2 \,\text{kPa}$$

Then new

$$\rho_1 = \frac{p_1}{RT_1}$$

$$= \frac{92.2 \times 10^3}{287 \times 281.4}$$

$$= 1.14 \, \text{kg/m}^3$$

and new

$$C_a = \frac{8}{1.14 \times 0.0574}$$

$$= 122.25 \, \text{m/s}$$

Repeat

$$\frac{C_1^2}{2C_p} = \frac{122.25^2}{2 \times 1005}$$

$$= 7.43 \, \text{K}$$

$$T_1 = 288 - 7.43$$

$$= 280.57 \, \text{K}$$

$$p_1 = 10^5 \left(\frac{280.57}{288}\right)^{3.5}$$

$$= 91.25 \, \text{kPa}$$

$$\rho_1 = \frac{91.25 \times 10^3}{287 \times 280.56}$$

$$= 1.13 \, \text{kg/m}^3$$

Further iterations are unnecessary and the value $\rho_1 = 1.13 \, \text{kg/m}^3$ may be taken as the inlet density and $C_a = C_1 = 122.25 \, \text{m/s}$ as the inlet velocity.

At the *tip*

$$U_1 = \frac{2\pi N r_t}{60} = \frac{2 \times \pi \times 16\,200 \times 0.15}{60}$$

$$= 254.5 \, \text{m/s}$$

Blade angle

$$\beta_{1t} = \tan^{-1}\left(\frac{U_1}{C_a}\right)$$

$$= \tan^{-1}\left(\frac{254.5}{122.25}\right)$$

$$= 64.34°$$

At the *hub*

$$U_1 = \frac{2 \times \pi \times 16\,200 \times 0.065}{60}$$

$$= 110.3 \, \text{m/s}$$

Blade angle

$$\beta_{1h} = \tan^{-1} \left(\frac{110.3}{122.25} \right)$$

$$= 42.05°$$

The Mach number required is that based on the relative velocity at the eye tip. Acoustic velocity

$$a_1 = (\gamma R T_1)^{1/2}$$

$$= (1.4 \times 287 \times 280.57)^{1/2}$$

$$= 335.8 \, \text{m/s}$$

From the inlet velocity triangle

$$W_1 = (C_a^2 + U_1^2)^{1/2}$$

$$= (122.25^2 + 254.5^2)^{1/2}$$

$$= 282.3 \, \text{m/s}$$

$$\text{Relative Mach number} = W_1/a_1$$

$$= \frac{282.3}{335.8}$$

$$\underline{M_{1,\text{rel}} = 0.841}$$

Exercise 4.2 Using Fig. 4.2, since there is no slip the blade angle at impeller exit equals the relative velocity vector angle ($\beta_2' = \beta_2$).

$$U_2 = \frac{\pi N D_2}{60}$$

$$= \frac{\pi \times 20\,000 \times 0.48}{60}$$

$$= 503 \, \text{m/s}$$

$$C_{x2} = U_2 - C_{r2} \tan \beta_2$$

$$= 503 - 122 \tan 26.5°$$

$$= 442 \, \text{m/s}$$

From Eq. (1.25) the power per unit mass flow is

$$W/m = U_2 C_{x2} - U_1 C_{x1}$$

But C_{x1} is zero, therefore

$$W/m = 503 \times 442$$

and

$$m = W/222.3 \times 10^3 \text{ m}^3/\text{s}$$

Theoretical power = Power available × Mechanical efficiency

$$W = 580 \times 10^3 \times 0.95$$

$$= 551 \text{ kW}$$

Hence

$$m = 551/222.3$$

Expected mass flow = 2.47 kg/s

Using the continuity equation, the mass flow rate through the annulus of the eye is

$$m = \rho_1 A_1 C_1$$

$$= \rho_1 \pi (r_t^2 - r_h^2) C_1$$

$$= \rho_1 \pi r_h^2 \left(\frac{r_t^2}{r_h^2} - 1 \right) C_1$$

Assuming incompressible flow at inlet

$$\rho_1 = \rho_{01}$$

$$= \frac{p_{01}}{R T_{01}}$$

$$= \frac{101.3 \times 10^3}{287 \times 288}$$

$$= 1.226 \text{ kg/m}^3$$

Thus

$$r_h^2 = \frac{2.47}{\pi \times 1.226 \times (3.333^2 - 1) \times 95}$$

$$= 6.677 \times 10^{-4} \text{ m}$$

$$r_h = 25.8 \text{ mm}$$

Hub diameter = 51.7 mm

$$r_t = 51.7/0.3$$

$$\text{Tip diameter} = 172 \text{ mm}$$

From Eq. (4.10), compressor total-to-total efficiency is

$$\eta_c = \frac{h_{03ss} - h_{01}}{h_{03} - h_{01}}$$

$$= \frac{C_p T_{01}[(T_{03ss}/T_{01}) - 1]}{W/m}$$

But

$$\frac{T_{03ss}}{T_{01}} = \left(\frac{p_{03}}{p_{01}}\right)^{(\gamma - 1)/\gamma}$$

Therefore

$$\eta_c = \frac{1005 \times 288(5.5^{0.286} - 1)}{222.3 \times 10^3}$$

$$\underline{\text{Efficiency} = 0.818}$$

Exercise 4.3 The absolute Mach number of the air leaving the impeller vanes is

$$M_2 = C_2/a_2 \qquad (a_2 = \text{acoustic velocity at impeller tip})$$

$$= \frac{C_2}{(\gamma R T_2)^{1/2}} \qquad (T_2 = \text{static temperature at impeller tip})$$

The problem therefore requires C_2 and T_2 to be evaluated.

Slip factor

$$\sigma_s = \frac{C'_{x2}}{C_{x2}}$$

$$C'_{x2} = \sigma_s U_2 \qquad \text{since } C_{x2} = U_2 \text{ for zero slip}$$

$$= 0.9 \times 365$$

$$= 328.5 \text{ m/s}$$

From the velocity triangles of Fig. 2.5,

$$C_2^2 = C_{r2}^2 + (C'_{x2})^2$$

$$= 30^2 + 328.5^2$$

$$= 1.088 \times 10^5 \text{ m}^2/\text{s}^2$$

With zero inlet whirl Eq. (4.2) gives

$$W/m = \sigma_s C_{x2} U_2$$

$$= \sigma_s U_2^2$$

$$= C_p (T_{02} - T_{01})$$

Thus

$$T_{02} = T_{01} + \frac{\sigma_s U_2^2}{C_p}$$

$$= 288 + \frac{0.9 \times 365^2}{1005}$$

$$= 407.3 \text{ K}$$

Also

$$T_2 = T_{02} - \frac{C_2^2}{2C_p}$$

$$= 407.3 - \frac{108\,800}{2 \times 1005}$$

$$= 353.2 \text{ K}$$

Hence

$$M_2 = \left(\frac{1.088 \times 10^5}{1.4 \times 287 \times 353.2} \right)^{1/2}$$

$$= 0.876$$

At impeller outlet $m = \rho_2 A_2 C_{r2}$ where $\rho_2 = p_2/RT_2$ and note that these are static properties. The static pressure p_2 is found by first solving for the impeller total pressure ratio p_{02}/p_{01}, and then relating p_2 to p_{02} by means of the isentropic pressure–temperature relationships.

From Eqs (4.10) and (4.11) rearranged for flow in the impeller:

$$\frac{p_{02}}{p_{01}} = \left[1 + \eta_c \left(\frac{T_{02}}{T_{01}} - 1 \right) \right]^{\gamma/(\gamma - 1)}$$

$$= \left[1 + 0.9 \left(\frac{407.3}{288} - 1 \right) \right]^{3.5}$$

$$= 3.03$$

Now

$$\frac{p_2}{p_{02}} = \frac{T_2}{T_{02}} \qquad \text{from isentropic relationships at a point}$$

$$\frac{p_2}{p_{02}} = \left(\frac{353.2}{407.3}\right)^{3.5}$$

$$= 0.607$$

Therefore

$$p_2 = \left(\frac{p_2}{p_{02}}\right)\left(\frac{p_{02}}{p_{01}}\right)p_{01}$$

$$= 0.607 \times 3.03 \times 101.3 \times 10^5$$

$$= 186.3\,\text{kPa}$$

$$\rho_2 = \frac{186.3 \times 10^3}{287 \times 353.2}$$

$$= 1.84\,\text{kg/m}^2$$

Mass flow

$$m = 1.84 \times 0.093 \times 30$$

$$= 5.13\,\text{kg/s}$$

Exercise 4.4 For radial vanes the slip factor may be found from the Stanitz equation (Eq. (2.9)) with $\beta_2 = 90°$.

Slip factor

$$\sigma_s = 1 - \frac{0.63\pi}{Z}$$

$$= 1 - \frac{0.63 \times \pi}{17}$$

$$= 0.884$$

Impeller tip speed

$$U_2 = \frac{\pi N D_2}{60}$$

$$= \frac{\pi \times 46\,000 \times 0.165}{60}$$

$$= 397.4\,\text{m/s}$$

$$W/m = \sigma_s U_2 C_{x2} - U_1 C_{x1}$$

But $C_{x1} = $ zero and C_{x2} equals U_2, the whirl velocity at impeller tip without slip. So

$$W = 0.6 \times 0.884 \times 397.4^2$$

Theoretical power $= 83.76\,\text{kW}$

(a) $r_t - r_h = 25$ mm and $(r_h + r_t)/2 = 31.75$ mm. Solving for r_h gives

$$r_h = \frac{(2 \times 31.75) - 25}{2}$$

$$= 19.25 \text{ mm}$$

$$r_t = 44.25 \text{ mm}$$

At the eye

$$\rho_1 = \frac{p_1}{RT_1}$$

$$= \frac{93 \times 10^3}{287 \times 293}$$

$$= 1.106 \text{ kg/m}^3$$

Mass flow

$$m = \rho_1 A_1 C_a \qquad \text{since } C_{x1} = 0$$

$$= \rho_1 A_1 C_1$$

$$C_1 = \frac{0.6}{1.107 \times \pi(0.044\,25^2 - 0.019\,25^2)}$$

$$= 108.7 \text{ m/s}$$

From the velocity triangles of Fig. 4.2

$$\tan \beta_1 = U_1/C_1$$

At the mean radius

$$U_1 = \frac{\pi N D_m}{60}$$

$$= 152.9 \text{ m/s}$$

$$\beta_1 = \tan^{-1}\left(\frac{152.9}{108.7}\right)$$

Inlet blade angle $= 54.6°$

(b)
$$h_{01} = h_1 + \frac{C_1^2}{2}$$

$$= C_p T_1 + \frac{C_1^2}{2}$$

$$= (1005 \times 293) + \frac{108.7^2}{2}$$

$$= 3.004 \times 10^5 \text{ N m/kg}$$

But from Eq. (4.5),

$$W/m = h_{02} - h_{01}$$

$$h_{02} = \frac{83\,760}{0.6} + 3.004 \times 10^5$$

$$= 4.4 \times 10^5 \, \text{N m/kg}$$

Therefore

$$T_{02} = \frac{4.4 \times 10^5}{C_p}$$

$$= \frac{4.4 \times 10^5}{1005}$$

$$= 437.8 \, \text{K}$$

(c) From Fig. 4.3 the impeller efficiency is

$$\eta_i = \frac{h_{02s} - h_{01}}{h_{02} - h_{01}}$$

$$C_p(T_{02s} - T_{01}) = \eta_i W/m$$

$$= \frac{0.9 \times 83.76 \times 10^3}{0.6}$$

$$= 125.64 \times 10^3$$

$$T_{02s} = \frac{125.64 \times 10^3}{C_p} + T_1 + \frac{C_1^2}{2C_p}$$

$$= \frac{125.64 \times 10^3}{1005} + 293 + \frac{108.7^2}{2 \times 1005}$$

$$= 423.9 \, \text{K}$$

From isentropic relationships

$$\frac{p_{02}}{p_{01}} = \left(\frac{T_{02s}}{T_{01}}\right)^{\gamma/(\gamma-1)}$$

Now

$$T_{01} = h_{01}/C_p$$

$$= \frac{300\,400}{1005}$$

$$= 298.9 \, \text{K}$$

$$\frac{p_1}{p_{01}} = \left(\frac{T_1}{T_{01}}\right)^{\gamma/(\gamma-1)}$$

$$p_{01} = 93\left(\frac{298.9}{293}\right)^{3.5}$$

$$= 99.7\,\text{kPa}$$

Therefore

$$p_{02} = 99.7\left(\frac{423.9}{298.9}\right)^{3.5}$$

$$= 338.67\,\text{kPa}$$

Exercise 4.5 (a) Using Eq. (4.22) the appropriate known data are substituted, noting that all conditions apply at the eye tip or shroud.

$$k = 1 - \left(\frac{r}{R}\right)^2$$

$$= 1 - 0.38^2$$

$$= 0.8556$$

The LHS of Eq. (4.22) becomes

$$\text{LHS} = \frac{4\omega^2}{\pi \times 0.8556 \times 1.4 \times 101.3 \times 10^3 \times (1.4 \times 287 \times 288)^{1/2}}$$

$$= 3.08 \times 10^{-8}\omega^2$$

and the RHS of Eq. (4.22) becomes

$$\text{RHS} = \frac{0.97^3 \sin^2 \beta_1 \cos \beta_1}{[1 + 0.5(1.4 - 1)0.95^2 \cos^2 \beta_1]^4}$$

$$= \frac{0.9127 \sin^2 \beta_1 \cos \beta_1}{(1 + 0.1805 \cos^2 \beta_1)^4}$$

Choosing values of β_1 the RHS of Eq. (4.22) is evaluated:

β_1(deg)	10	20	30	40	50	55	59	60	61	65	70	80
RHS	0.014	0.055	0.119	0.193	0.258	0.279	0.286	0.287	0.287	0.279	0.253	0.15

The results are plotted in Fig. 4.11 from which it is deduced that the RHS reaches a maximum at $\beta_1 = 60°$.

Equating the LHS and RHS with $\beta_1 = 60°$,

$$3.08 \times 10^{-8}\omega^2 = \frac{0.9127 \sin^2 60° \cos 60°}{(1 + 0.1805 \cos^2 60°)^4}$$

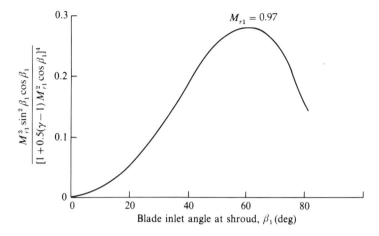

Figure 4.11 Mass flow parameter versus blade inlet angle at the shroud

$$\omega^2 = \frac{0.2869}{3.08 \times 10^{-8}}$$

$$= 9\,313\,987$$

$$\omega = 3051 \text{ rad/s}$$

$$\underline{\text{Optimum speed} = 29\,143 \text{ rpm}}$$

(b) The eye tip diameter is found from the continuity equation for flow into the eye. Once again all equations apply at the tip radius.

$$m = \rho_1 A_1 C_1 \qquad \text{since } C_1 = C_a \text{ (no whirl)}$$

Expressing the area in terms of the radius ratio and tip radius

$$m = \pi \rho_1 k R^2 C_1$$

It is necessary to determine C_1 and ρ_1.

To find C_1, from the inlet velocity triangle of Fig. 4.2 at the eye tip radius

$$W_1 = C_1/\cos\beta_1$$

But

$$M_{r1} = \frac{W_1}{a_1} \qquad \text{and} \qquad M_1 = \frac{C_1}{a_1}$$

Therefore

$$M_1 = M_{r1}\cos\beta_1$$

$$= 0.97\cos 60°$$

$$= 0.485$$

Now

$$\frac{T_{01}}{T_1} = 1 + \frac{C_1^2}{2C_p T_1}$$

Putting

$$C_p T_1 = \frac{\gamma R T_1}{\gamma - 1}$$

then

$$\frac{T_{01}}{T_1} = 1 + \frac{(\gamma - 1)C_1^2}{2\gamma R T_1}$$

and putting

$$a_1^2 = \gamma R T_1$$

then

$$\frac{T_{01}}{T_1} = 1 + \frac{(\gamma - 1)M_1^2}{2}$$

$$= 1 + \frac{0.4 \times 0.485^2}{2}$$

$$= 1.047$$

Whence

$$T_1 = \frac{288}{1.047}$$

$$= 275 \text{ K}$$

Then

$$C_1 = M_1 (\gamma R T_1)^{1/2}$$

$$= 0.485(1.4 \times 287 \times 275)^{1/2}$$

$$= 161 \text{ m/s}$$

From isentropic relationships at a point

$$\frac{\rho_1}{\rho_{01}} = \left(\frac{T_1}{T_{01}}\right)^{1/(\gamma - 1)}$$

$$\rho_1 = \frac{p_{01}}{R T_{01}}\left(\frac{T_1}{T_{01}}\right)^{1/(\gamma - 1)}$$

$$= \frac{101.3 \times 10^3}{287 \times 288}\left(\frac{1}{1.04}\right)^{2.5}$$

$$= 1.11 \text{ kg/m}^3$$

Therefore

$$R^2 = \frac{m}{\pi \rho_1 k C_1}$$

$$= \frac{4}{\pi \times 1.11 \times 0.8556 \times 161}$$

$$= 0.008\,33$$

$$R = 0.0913 \, \text{m}$$

Eye tip diameter $= 183 \, \text{mm}$

Exercise 4.6 From Eq. (4.11) the stagnation temperature difference is

$$T_{03} - T_{01} = \frac{T_{01}}{\eta_c} \left[\left(\frac{p_{03}}{p_{01}} \right)^{0.286} - 1 \right]$$

$$= \frac{288}{0.82} (3.8^{0.286} - 1)$$

$$= 163.3 \, \text{m/s}$$

Now

$$T_{03} - T_{01} = \frac{\psi \sigma_s U_2^2}{C_p} \qquad \text{(from Eq. (4.9))}$$

Therefore

$$U_2^2 = \frac{163.3 \times 1005}{1.04 \times 0.92}$$

$$= 171\,526$$

$$U_2 = 414.15 \, \text{m/s}$$

Also

$$U_2 = \frac{\pi N D_2}{60}$$

$$D_2 = \frac{414.15 \times 60}{\pi \times 12\,000}$$

Impeller outer diameter $= 0.659 \, \text{m}$

The overall loss is proportional to $(1 - \eta_c) = (1 - 0.82)$. Half of the overall loss is therefore $0.5(1 - 0.82) = 0.09$ and therefore the effective efficiency of the impeller in compressing from p_{01} to p_{02} is $(1 - 0.09) = 0.91$.

From Eq. (4.11) after rearranging the subscripts

$$\frac{p_{02}}{p_{01}} = \left(1 + \frac{\eta_i(T_{02} - T_{01})}{T_{01}}\right)^{\gamma/(\gamma-1)}$$

$$= \left(1 + \frac{0.91 \times 163.3}{288}\right)^{3.5} \quad \text{since } T_{02} = T_{03}$$

$$= 4.29$$

$$T_{02} = 163.3 + 288$$

$$= 451.3 \text{ K}$$

$$T_2 = T_{02} - \frac{C_2^2}{2C_p} \quad \text{and} \quad M_2 = \frac{C_2}{(\gamma R T_2)^{1/2}}$$

Therefore substituting for C_2^2

$$T_2 = 451.3 - \frac{(1)^2(1.4 \times 287 T_2)}{2 \times 1005}$$

$$T_2 = \frac{451.3}{1 + 0.2}$$

$$= 376.1 \text{ K}$$

$$\frac{p_2}{p_{02}} = \left(\frac{T_2}{T_{02}}\right)^{\gamma/(\gamma-1)}$$

$$= \left(\frac{376.1}{451.3}\right)^{3.5}$$

$$= 0.5282$$

$$\frac{p_2}{p_{01}} = \left(\frac{p_2}{p_{02}}\right)\left(\frac{p_{02}}{p_{01}}\right)$$

$$= 0.528 \times 4.29$$

$$= 2.266$$

and

$$p_2 = 2.266 \times 101.3$$

$$= 229.58 \text{ kPa}$$

$$\rho_2 = \frac{p_2}{R T_2}$$

$$= \frac{229.58 \times 10^3}{287 \times 376.1}$$

$$= 2.127 \text{ kg/m}^3$$

To find the flow velocity normal to the periphery of the impeller

$$C_{x2} = \sigma_s U_2$$
$$= 0.92 \times 414.15$$
$$= 381 \text{ m/s}$$
$$C_2^2 = M_1^2 \gamma R T_2$$
$$= (1)^2 \times (1.4 \times 287 \times 376.1)$$
$$= 151\,117 \text{ m}^2/\text{s}^2$$

From the velocity triangles of Fig. 4.2

$$C_{r2}^2 = C_2^2 - C_{x2}'^2$$
$$= 151\,117 - 145\,161$$
$$= 5956 \text{ m}^2/\text{s}$$
$$C_{r2} = 77.17 \text{ m/s}$$

From the continuity equation the area

$$A_2 = \frac{m}{\rho_2 C_{r2}}$$
$$= \frac{8}{2.127 \times 77.17}$$
$$= 0.0487 \text{ m}^2$$

$$\text{Depth of impeller} = \frac{A_2}{\pi D_2}$$
$$= \frac{0.0487}{\pi \times 0.659}$$
$$= 23.5 \text{ mm}$$

Exercise 4.7 Figure 2.18 shows the absolute velocity vector C at angle α' to the radial direction. When the gas flows into the diffuser passage, it must do so at this angle and hence α' is the inlet angle of the diffuser vanes:

$$\tan \alpha' = C_x/C_r$$

In the vaneless space between the impeller outlet and diffuser vanes the flow is that of a free vortex which at any radius requires that $C_x r = \text{const.}$

At the diffuser vane leading edge the radius is $(r_2 + 40) \text{ mm} = (329.5 + 40) = 369.5 \text{ mm}$ and

$$C_x = \frac{C_2 r_2}{r}$$

$$= \frac{(151\,117)^{1/2} \times 0.3295}{0.3695}$$

$$= 346.7\,\text{m/s}$$

To find the radial velocity C_r at the diffuser vane entry, start by assuming the value at the impeller exit, i.e. 77.17 m/s. Then

$$\frac{C^2}{2C_p} = \frac{C_r^2 + C_x^2}{2C_p}$$

$$= \frac{77.17^2 + 346.7^2}{2 \times 1005}$$

$$= 125.5\,\text{K}$$

Now if we assume that no losses occur across the vaneless space, the other half of the total losses takes place in the diffuser itself. Then p_{02} at the impeller tip equals the stagnation pressure at the diffuser vane inlet p_0. Therefore

$$p_0 = 4.29 p_{01} \qquad \text{(from exercise 4.6)}$$

$$T = T_0 - \frac{C^2}{2C_p}$$

$$= 451.3 - 125.5 \qquad (\text{since } T_0 = T_{02})$$

$$= 325.8$$

$$\frac{p}{p_{02}} = \left(\frac{T}{T_{02}}\right)^{\gamma/(\gamma-1)}$$

$$= \left(\frac{325.8}{451.3}\right)^{3.5}$$

$$= 0.32$$

Also

$$\frac{p}{p_{01}} = \left(\frac{p}{p_{02}}\right)\left(\frac{p_{02}}{p_{01}}\right)$$

$$= 0.32 \times 4.29$$

$$= 1.373$$

$$p = 1.373 \times 101.3$$

$$= 139\,\text{kPa}$$

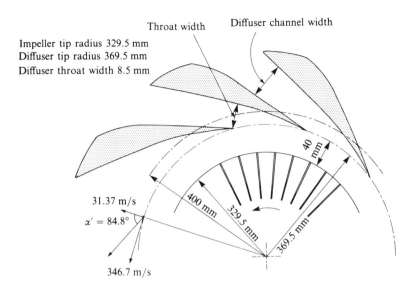

Figure 4.12 Flow from impeller to diffuser

$$\rho = \frac{p}{RT}$$

$$= \frac{139 \times 10^3}{287 \times 325.8}$$

$$= 1.487 \, \text{kg/m}^3$$

With reference to Fig. 4.12, the area of flow in the radial direction at radius 0.369 m is

$$A_r = 2\pi \times 0.369 \times \text{Depth of vane}$$

$$= 2\pi \times 0.369 \times 0.0235$$

$$= 0.0545 \, \text{m}^2$$

Therefore

$$C_r = \frac{m}{\rho A}$$

$$= \frac{4}{1.487 \times 0.0545}$$

$$= 49.4 \, \text{m/s}$$

Repeat the calculation using this new value of C_r.

$$C^2/(2C_p) = 61 \text{ K}$$

$$T = 390.3 \text{ K}$$

$$p/p_{02} = 0.601$$

$$p = 261.4 \text{ kPa}$$

$$\rho = 2.33 \text{ kg/m}^3$$

$$C_r = 31.6 \text{ m/s}$$

Repeat the calculation.

$$C^2/(2C_p) = 60.29 \text{ K}$$

$$T = 391 \text{ K}$$

$$p/p_{02} = 0.605$$

$$p = 263 \text{ kPa}$$

$$\rho = 2.34 \text{ kg/m}^3$$

$$C_r = 31.37 \text{ m/s}$$

No further iterations are necessary. Thus at the inlet to the vanes $C_r = 31.37 \text{ m/s}$. Then

$$\tan \alpha' = \frac{346.7}{31.37}$$

$$\alpha' = 84.83°$$

Moving to the radius at the diffuser throat, at the throat radius, 0.4 m,

$$C_x = \frac{(151\,117)^{1/2} \times 0.3295}{0.4}$$

$$= 320.2 \text{ m/s}$$

Start the iteration at the previous value of C_r at the diffuser inlet.

$$C^2/(2C_p) = 51.5 \text{ K}$$

$$T = 399.8 \text{ K}$$

$$p/p_{02} = 0.654$$

$$p = 284 \text{ kPa}$$

$$\rho = 2.47 \text{ kg/m}^3$$

Neglecting vane thickness

$$\text{Flow area} = 2\pi \times \text{Radius of diffuser throat} \times \text{Depth}$$

$$= 2\pi \times 0.4 \times 0.0235$$

$$A_r = 0.0591 \text{ m}^2$$

$$C_r = \frac{4}{2.47 \times 0.0591}$$

$$= 27.4 \text{ m/s}$$

Recalculate.

$$C^2/(2C_p) = 51.38 \text{ K}$$

$$T = 399.9 \text{ K}$$

It may be seen that there is no change in the new values so the radial velocity at the diffuser throat = 27.4 m/s.

The flow direction of the absolute velocity is given by α' at the throat:

$$\tan \alpha' = C_x/C_r$$

$$= \frac{320.2}{27.4}$$

$$\alpha' = 85.1°$$

With reference to Fig. 4.12

$$m = \rho AC = \rho A_r C_r \qquad (r \text{ means w.r.t. radial direction})$$

$$A = \frac{A_r C_r}{C}$$

But

$$C = (51.38 \times 2C_p)^{1/2}$$

$$= 321.4 \text{ m/s}$$

and

$$A_r \equiv 0.0235 \quad \text{(the vane depth)}$$

Therefore

$$A = \frac{0.0235 \times 27.4}{321.4}$$

$$= 0.002 \text{ m}^2$$

As we have 10 diffuser vanes, the width of each throat is

$$\text{Throat width} = \frac{A}{10 \times \text{Depth}}$$

$$= \frac{0.002}{10 \times 0.0235}$$

$$= 0.0085$$

$$= 8.5\,\text{mm}$$

Exercise 4.8 The velocity triangles of Fig. 4.2 may be used.

$$M_{1r} = W_1/a_1$$

$$T_1 = T_{01} - \frac{C_1^2}{2C_p}$$

$$= 288 - \frac{90^2}{2 \times 1005}$$

$$= 284\,\text{K}$$

Therefore

$$a_1 = (\gamma R T_1)^{1/2}$$

$$= (1.4 \times 287 \times 284)^{1/2}$$

$$= 337.8\,\text{m/s}$$

At the *tip*

$$U_1 = \frac{\pi N D_1}{60}$$

$$= \frac{12\,000 \times \pi \times 0.45}{60}$$

$$= 282.7\,\text{m/s}$$

$$W_1 = 0.75 \times 337.8$$

$$= 253.35\,\text{m/s}$$

Now from Fig. 4.7 at the inlet tip

$$W_1^2 = (U_1 - C_{x1})^2 + C_a^2 \quad \text{(at the tip)}$$

$$= U_1^2 - 2U_1 C_{x1} + C_{x1}^2 + C_1^2 - C_{x1}^2$$

$$= U_1^2 - 2U_1 C_{x1} + C_1^2$$

$$C_{x1} = \frac{282.7^2 + 90^2 - 253.35^2}{2 \times 282.7}$$

$$= 42.15 \, \text{m/s}$$

But

$$C_{x1} = C_1 \sin \alpha_1$$

Therefore

$$\alpha_1 = \sin^{-1}\left(\frac{C_{x1}}{C_1}\right)$$

$$= \sin^{-1}\left(\frac{42.15}{90}\right)$$

Blade angle at tip $= 27.9°$

The impeller total-to-total efficiency is given by

$$\eta_i = \frac{T_{02s} - T_{01}}{T_{02} - T_{01}}$$

$$= \left[\left(\frac{p_{02}}{p_{01}}\right)^{(\gamma-1)/\gamma} - 1\right] \Big/ \left(\frac{T_{02}}{T_{01}} - 1\right)$$

Now

$$W/m = C_p(T_{02} - T_{01})$$

$$= \sigma_s U_2 C_{x2} - U_1 C_{x1}$$

Therefore

$$\frac{T_{02}}{T_{01}} - 1 = \frac{\sigma_s U_2^2 - U_1 C_{x1}}{C_p T_{01}} \qquad \text{(since } U_2 = C_{x2}$$

Now

$$U_2 = \frac{\pi N D_2}{60}$$

$$= \frac{\pi \times 12\,000 \times 0.76}{60}$$

$$= 477.5 \, \text{m/s}$$

Hence

$$\frac{T_{02}}{T_{01}} = 1 + \frac{0.9 \times 477.5^2 - 282.7 \times 42.15}{288 \times 1005}$$

$$= 1 + 0.668$$

$$= 1.668$$

From the efficiency equation

$$p_{02}/p_{01} = [(1.668 - 1)0.9 + 1]^{3.5}$$
$$= 5.2$$

Now

$$\frac{p_2}{p_{01}} = \left(\frac{p_{02}}{p_{01}}\right)\left(\frac{p_2}{p_{02}}\right)$$
$$= \left(\frac{p_{02}}{p_{01}}\right)\left(\frac{T_2}{T_{02}}\right)^{\gamma/(\gamma-1)}$$

Also

$$T_{02} = T_2 + \frac{C_2^2}{2C_p}$$

Therefore

$$\frac{T_2}{T_{02}} = 1 - \frac{C_2^2}{2C_p T_{02}}$$
$$= 1 - \frac{C_{x2}'^2 + C_{r2}^2}{2C_p T_{02}}$$

But

$$C_{x2}' = \sigma_s U_2$$
$$= 0.9 \times 477.5$$
$$= 429.75 \text{ m/s}$$
$$T_{02} = 1.668 \times 288$$
$$= 480.3 \text{ K}$$

Substituting

$$\frac{T_2}{T_{02}} = 1 - \frac{429.75^2 + 50^2}{2 \times 1005 \times 480.3}$$
$$= 0.806$$
$$p_2/p_{01} = 5.2 \times 0.806^{3.5}$$
$$= 2.44$$

Impeller outlet static pressure = 265.8 kPa

Exercise 4.9 (a) Impeller tip speed

$$U_2 = \frac{\pi N D_2}{60}$$

$$= \frac{\pi \times 16\,000 \times 0.55}{60}$$

$$= 460.7 \, \text{m/s}$$

From Eq. (4.9)

$$T_{03} - T_{01} = \frac{\psi \sigma_s U_2^2}{1005} \quad \text{and} \quad T_{03} = T_{02}$$

Therefore

$$T_{02} - T_{01} = \frac{1.04 \times 0.9 \times 460.7^2}{1005}$$

$$= 197 \, \text{K}$$

$$\frac{T_{02}}{T_{01}} = \frac{197}{288} + 1$$

$$= 1.68$$

Now for the impeller

$$\frac{p_{02}}{p_{01}} = \left(1 + \frac{\eta_i (T_{02} - T_{01})}{288}\right)^{3.5}$$

$$= \left(1 + \frac{0.9 \times 197}{288}\right)^{3.5}$$

$$= 5.36$$

Then

$$\underline{p_{02} = 541.4 \, \text{kPa}}$$

and

$$T_{02} = 197 + 288$$

$$\underline{= 485 \, \text{K}}$$

(b) From

$$C'_{x2} = \sigma_s U_2$$

$$= 0.9 \times 460$$

$$= 414 \, \text{m/s}$$

Let $C_{r2} = 100 \, \text{m/s}$.

Outlet area normal to impeller periphery

$$A_2 = \pi D_2 \times \text{Impeller depth}$$

$$= \pi \times 0.55 \times 0.037$$

$$= 0.0639 \, \text{m}^2$$

From the outlet velocity triangle

$$\frac{C_2^2}{2C_p} = \frac{C_{r2}^2 + C_{x2}'^2}{2C_p}$$

$$= \frac{100^2 + 414^2}{2 \times 1005}$$

$$= 90.25 \text{ K}$$

Thus

$$T_2 = T_{02} - \frac{C_2^2}{2C_p}$$

$$= 485 - 90.25$$

$$= 394.75 \text{ K}$$

From isentropic relationships at a point

$$p_2 = p_{02} \left(\frac{T_2}{T_{02}} \right)^{\gamma/(\gamma-1)}$$

$$= 541.4 \left(\frac{394.75}{485} \right)^{3.5}$$

$$= 263.4 \text{ kPa}$$

Using the equation of state

$$\rho_2 = \frac{p_2}{RT_2}$$

$$= \frac{263.4}{287 \times 394.75}$$

$$= 2.33 \text{ kg/m}^3$$

The equation of continuity gives

$$C_{r2} = \frac{m}{A_2 \rho_2}$$

$$= \frac{15}{0.0639 \times 2.33}$$

Impeller outlet radial velocity $= 100.7 \text{ m/s}$

Impeller outlet Mach number $= C_2/a_2$

Now

$$a_2 = (\gamma R T_2)^{1/2}$$

$$= (1.4 \times 287 \times 394.75)^{1/2}$$
$$= 398.3 \text{ m/s}$$

and

$$C_2 = (90.25 \times 2 \times 1005)^{1/2}$$
$$= 425.9 \text{ m/s}$$

$$M_2 = \frac{425.9}{398.3}$$

$$= 1.07$$

At the outlet velocity triangle

$$\cos \alpha_2 = C_{r2}/C_2$$

$$= \frac{100}{425.9}$$

$$\alpha_2 = 76.4°$$

(c) Assuming free vortex flow in the vaneless space and denoting conditions at the diffuser vane tip without a subscript,

$$C_x = \frac{C_{x2} r_2}{r}$$

$$= \frac{414 \times 0.275}{0.317}$$

$$= 359.1 \text{ m/s}$$

Choose as a first try $C_r = 100 \text{ m/s}$. Then

$$\frac{C^2}{2C_p} = \frac{100^2 + 359.1^2}{2 \times 1005}$$

$$= 69.13 \text{ K}$$

Following the same procedure as in part (b)

$$T = 485 - 69.13 \qquad (T = T_{02} \text{ in vaneless space})$$
$$= 415.9 \text{ K}$$

$$p = p_{02} \left(\frac{T}{T_{02}} \right)^{3.5}$$

$$= 541.4 \left(\frac{415.9}{485} \right)^{3.5}$$

$$= 316.3 \text{ kPa}$$

$$\rho_2 = \frac{316.3}{287 \times 415.9}$$

$$= 2.65$$

$$A = 2\pi r \times \text{Depth of vanes}$$

$$= 2 \times \pi \times 0.317 \times 0.037$$

$$= 0.0737 \,\text{m}^2$$

$$C_r = \frac{15}{2.65 \times 0.0737}$$

$$= 76.8 \,\text{m/s}$$

Next try $C_r = 76.8 \,\text{m/s}$.

$$\frac{C^2}{2C_p} = \frac{76.8^2 + 359.1^2}{2 \times 1005}$$

$$= 67.09 \,\text{m}^2/\text{s}^2$$

$$T = 417.9 \,\text{K}$$

$$p = 321.7 \,\text{kPa}$$

$$\rho_2 = 2.68 \,\text{kg/m}^3$$

$$C_r = 75.9 \,\text{m/s}$$

Therefore there is little change in C_r and thus this is the radial velocity at the diffuser vane tip.

At the vane

$$\tan \alpha' = C_x / C_r$$

$$= \frac{359.1}{75.9}$$

Vane angle

$$\alpha' = 78°$$

Mach number at vane

$$M = C/a$$

$$= \left(\frac{67.09 \times 2 \times 1005}{1.4 \times 287 \times 417.9}\right)^{1/2}$$

Diffuser vane Mach number $= 0.896$

Exercise 4.10 (a) Mechanical efficiency $= \dfrac{\text{Power transferred to air}}{\text{Power supplied to shaft}}$

$$\text{Shaft power} = \frac{W}{\eta_m}$$

Now for a radial vaned impeller the Stanitz slip factor equation is used.

$$\sigma_s = 1 - \frac{0.63\pi}{Z}$$

$$= \frac{0.63\pi}{19}$$

$$= 0.8958$$

Now

$$
\begin{aligned}
W/m &= h_{03} - h_{01} \\
&= \psi\sigma_s U_2 C_{x2} \qquad \text{since } C_{x1} = 0 \\
&= \psi\sigma_s U_2^2 \qquad \text{since } U_2 = C_{x2} \\
W &= 1.04 \times 0.8958 \times 525^2 \times 2 \\
&= 513.56 \, \text{kW}
\end{aligned}
$$

$$
\begin{aligned}
\text{Shaft power} &= \frac{513.56}{0.96} \\
&= 535 \, \text{kW}
\end{aligned}
$$

(b) From Eq. (4.11)

$$
\begin{aligned}
\frac{p_{03}}{p_{01}} &= \left(1 + \frac{\eta_c \psi\sigma_s U_2^2}{C_p T_{01}}\right)^{\gamma/(\gamma-1)} \\
&= \left(1 + \frac{0.81 \times 1.04 \times 0.8958 \times 525^2}{1005 \times 290}\right)^{3.5} \\
&= 6.59
\end{aligned}
$$

Diffuser outlet stagnation pressure

$$
\begin{aligned}
p_{03} &= 6.59 \times 101 \\
&= 665 \, \text{kPa}
\end{aligned}
$$

$$\frac{p_3}{p_{03}} = \left(\frac{T_3}{T_{03}}\right)^{\gamma/(\gamma-1)}$$

But

$$T_{03} = T_3 + \frac{C_3^2}{2C_p}$$

$$\frac{T_3}{T_{03}} = 1 - \frac{95^2}{2 \times 1005 \times T_{03}}$$

Also

$$T_{03} = \frac{W}{C_p m} + T_{01}$$

$$= \frac{513.56 \times 10^3}{2 \times 1005} + 290$$

$$= 545.5 \text{ K}$$

Thus

$$\frac{T_3}{T_{03}} = 1 - \frac{95^2}{2 \times 1005 \times 545.5}$$

$$= 0.9917$$

Substituting

$$p_3 = 665(0.9917)^{3.5}$$

$$= 646 \text{ kPa}$$

Diffuser outlet static pressure $= 646 \text{ kPa}$

(c) At the impeller outlet

$$T_{02} = T_2 + \frac{C_2^2}{2C_p}$$

or

$$T_{03} = T_2 + \frac{C_2^2}{2C_p} \qquad \text{since } T_{02} = T_{03}$$

Rearranging

$$C_2^2 = 2C_p[(T_{03} - T_{01}) + (T_{01} - T_2)]$$

From the equation for reaction

$$T_2 - T_1 = 0.5(T_3 - T_1)$$

and

$$T_3 - T_1 = (T_{03} - T_{01}) + \left(\frac{C_1^2 - C_3^2}{2C_p}\right)$$

$$= \frac{W}{C_p m} + \left(\frac{C_1^2 - C_3^2}{2C_p}\right)$$

$$= \frac{513.56 \times 10^3}{1005 \times 2} + \left(\frac{160^2 - 95^2}{2 \times 1005} \right)$$

$$= 255.5 + 8.25$$

$$= 263.7 \, \text{K}$$

Substituting

$$T_2 - T_1 = 0.5 \times 263.7$$

$$= 131.87 \, \text{K}$$

Now

$$T_2 = T_{01} - \frac{C_1^2}{2C_p} + (T_2 - T_1)$$

$$= 290 - \frac{160^2}{2 \times 1005} + 131.87$$

$$= 290 - 12.74 + 131.87$$

$$= 409.1 \, \text{K}$$

Hence

$$C_2 = [2 \times 1005(255.5 + 290 - 409.1)]^{1/2}$$

$$= 523.6 \, \text{m/s}$$

Acoustic velocity

$$a_2 = (\gamma R T_2)^{1/2}$$

$$= (1.4 \times 287 \times 409.1)^{1/2}$$

$$= 405.4$$

Mach number at impeller outlet

$$M_2 = C_2/a_2$$

$$= \frac{523.6}{405.4}$$

$$= 1.29$$

Radial velocity at impeller outlet

$$C_{r2}^2 = C_2^2 - C_{x2}'^2$$

$$= 523.6^2 - (\sigma_s U_2)^2$$

$$= 523.6^2 - (0.8958 \times 525)^2$$

$$= 523.6^2 - 470.3^2$$

$$C_{r2} = 230.2 \, \text{m/s}$$

Knowing the diffuser efficiency, we can work back from the diffuser outlet to the impeller outlet. Diffuser efficiency

$$\eta_D = \frac{h_{3s} - h_2}{h_3 - h_2}$$

and this relates the isentropic enthalpy increase to the actual enthalpy increase. Rearranging

$$\eta_D = T_2\left(\frac{T_{3s}}{T_2} - 1\right) \bigg/ \left(T_3 - T_2\right)$$

$$= \left[T_2\left(\frac{p_3}{p_2}\right)^{(\gamma-1)/\gamma} - 1\right] \bigg/ \left(T_3 - T_2\right)$$

whence

$$\frac{p_3}{p_2} = \left(1 + \eta_D \frac{T_3 - T_2}{T_2}\right)^{\gamma/(\gamma-1)}$$

$$= \left(1 + \frac{0.83 \times 131.87}{409.1}\right)^{3.5} \qquad \text{since } (T_2 - T_1) = (T_3 - T_2)$$

$$= 2.29$$

$$p_2 = \frac{646}{2.29}$$

$$= 282.1\,\text{kPa}$$

Using isentropic relationships at a point

$$\frac{p_{02}}{p_2} = \left(\frac{T_{02}}{T_2}\right)^{\gamma/(\gamma-1)}$$

$$p_{02} = 282.1\left(\frac{545.5}{409.1}\right)^{3.5}$$

$$= 772.3\,\text{kPa}$$

(d) $$\eta_i = T_{01}\left[\left(\frac{p_{02}}{p_{01}}\right)^{(\gamma-1)/\gamma} - 1\right] \bigg/ \left(T_{03} - T_{01}\right)$$

$$= 290\left[\left(\frac{772.3}{101}\right)^{0.286} - 1\right] \bigg/ \left(545.5 - 290\right)$$

$$= \frac{228.9}{255.5}$$

$$= 0.895$$

(e) Assume that $C_x \gg C_r$ and if the gas moves in a free vortex then

$$C_{x3}r_3 = C_{x2}r_2$$

Now since $C_{x3} \gg C_{r3}$, then $C_{x3} \approx C_3$ and

$$\frac{r_2}{r_3} = \frac{C_{x3}}{C_{x2}}$$

$$= \frac{C_3}{\sigma_s U_2} \qquad \text{since } C_{x2} \equiv C'_{x2}$$

$$= \frac{95}{0.8958 \times 525}$$

$$\underline{r_2/r_3 = 0.202}$$

(f)

$$\rho_2 = \frac{p_2}{RT_2}$$

$$= \frac{282.1 \times 10^3}{287 \times 409.1}$$

$$= 2.402 \,\text{kg/m}^3$$

Mass flow rate at impeller outlet

$$m = \rho_2 A_2 C_{r2}$$
$$= 2\pi \rho_2 r_2 C_{r2} b_2$$

But

$$U_2 = \frac{\pi N D_2}{60} \qquad \text{where } N \text{ is in rpm}$$

$$= \frac{\pi N m}{\rho_2 \pi C_{r2} b_2 \times 60}$$

and

$$N = \frac{525 \times 2.402 \times 230.2 \times 0.006 \times 60}{2}$$

$$\underline{= 52\,253 \,\text{rpm}}$$

AXIAL FLOW COMPRESSORS AND FANS

5.1 INTRODUCTION

Because of a lack of knowledge of the aerodynamic behaviour of axial flow compressor blades, the initial simple concept of using a reversed turbine as an axial flow compressor resulted in compressor efficiencies of less than 40 per cent. Some early gas turbines did use axial flow compressors but these gave turbine efficiencies of the order of 55 per cent, and it was only with the development of aeroplanes, and research into the aerodynamic behaviour of wing sections, that blade design for axial flow compressors became established. Some early investigators suggested that efficiencies of 90 per cent would be reached, and this is now indeed the case, this figure surpassing the maximum centrifugal compressor efficiency by about 4 per cent. However, a penalty has to be paid for such high efficiencies and, with the axial flow compressor, the performance is very sensitive to its mass flow rate at the design point. Any deviation from the design condition causes the efficiency to drop off drastically. Thus the axial flow compressor is ideal for constant-load applications such as in aircraft gas turbine engines. They are also to be found in fossil fuel power stations where gas turbines are used for topping up the station output when normal peak loads are exceeded.

The simple expedient of using a turbine to operate as a compressor ran into difficulties because of the nature of the air flow in the two cases. In the turbine, the blades form a converging passage, the area at inlet being greater than at outlet, with the fluid being accelerated in the passage. In compressors, the fluid is diffused with a pressure gain taking place through a blade passage of increasing cross-sectional area. While a fluid can be accelerated over a wide

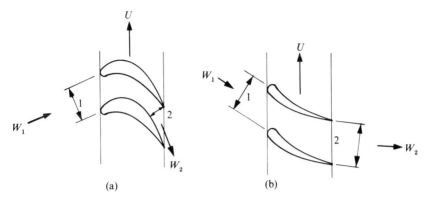

Figure 5.1 Compressor and turbine blade passages: (a) turbine; (b) compressor

range with high efficiency, the process of diffusion cannot be carried out so rapidly due to the onset of separation on the suction side of the blades, and consequent stalling. This is similar to the included angle of a diffuser being too great and separation taking place along the diffuser walls. Typical blade sections are shown in Fig. 5.1, where it will be noted that the angular turning of the relative velocity vector is much greater in the turbine than in the compressor. The maximum rate of efficient diffusion within the blade rows is equivalent to a cone angle of about 7° or 8°.

In studying the flow of the fluid through an axial compressor, it is usual to consider the changes taking place through a compressor stage. A stage consists of a row of moving blades attached to the periphery of a rotor hub followed by a row of fixed blades attached to the walls of the outer casing. The compressor is made up of a number of such stages to give an overall pressure ratio from inlet to outlet. Figure 5.2 illustrates a few compressor stages.

It will be seen that at the inlet to the compressor, an extra row of fixed vanes, called inlet guide vanes, are fitted. These do not form part of the stage but are solely to guide the air at the correct angle onto the first row of moving blades. The height of the blades is also seen to decrease as the fluid moves through the compressor. This is so that a constant axial velocity through the compressor is maintained as the density increases from the low- to high-pressure regions. A constant axial velocity is convenient from the point of view of design but is by no means a requirement. The analysis for flow through the stage will first be described in terms of two-dimensional flow. The flow through the stage is assumed to take place at a mean blade height where the blade peripheral velocities at inlet and outlet are the same, there being no flow in the radial direction. Whirl components of velocity will exist in the direction of blade motion.

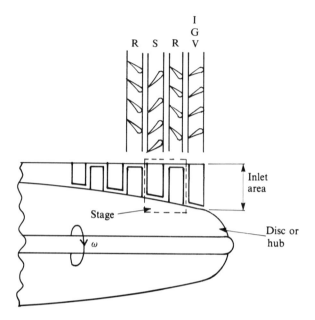

Figure 5.2 An axial compressor stage

5.2 COMPRESSOR STAGE

The rotor and stator rows of a stage are shown in Fig. 5.3. Note that all angles are referred to the axial velocity vector C_a.

Air exits from the previous row of stator blades at angle α_1 with absolute velocity C_1. The rotor row has tangential velocity U, and combining the two velocity vectors gives the relative inlet velocity vector W_1 at angle β_1. At rotor row outlet the velocity triangles are similar to those drawn for the axial flow pump, and the absolute velocity vector C_2 moves into the stator row where the flow direction is changed to α_3 with absolute velocity C_3. The diagrams have been drawn showing a large gap between the rotor and stator blades; this is for clarity. In practice, the clearance between rotor and stator rows is small.

If the following stage is the same as the preceding one, the stage is said to be normal. For a normal stage $C_1 = C_3$ and $\alpha_1 = \alpha_3$. W_2 is less than W_1, showing that diffusion of the relative velocity has taken place with some static pressure rise across the rotor blades. The air is turned towards the axial direction by the blade camber and the effective flow area is increased from inlet to outlet, thus causing diffusion to take place. Similar diffusion of the absolute velocity takes place in the stator, where the absolute velocity vector is again turned towards the axial direction and a further static pressure rise occurs.

The energy given to the air per unit mass flow rate is given by Euler's

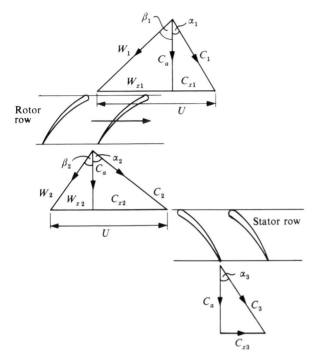

Figure 5.3 Velocity triangles for an axial flow compressor stage

equation (Eq. (1.25))

$$W/m = U_2 C_{x2} - U_1 C_{x1} \tag{5.1}$$

or

$$E = (U_2 C_{x2} - U_1 C_{x1})/g$$

But from the velocity triangles, and noting that C_a is constant through the stage, and $U_1 = U_2 = U$,

$$C_{x2} = U - C_a \tan \beta_2$$

and

$$C_{x1} = U - C_a \tan \beta_1$$

Thus

$$C_{x2} - C_{x1} = C_a(\tan \beta_1 - \tan \beta_2)$$

and

$$E = U C_a(\tan \beta_1 - \tan \beta_2)/g \quad (\text{W/(N/s)}) \tag{5.2}$$

The energy transfer may also be written in terms of the absolute velocity flow angles

$$E = U C_a(\tan \alpha_2 - \tan \alpha_1)/g \quad (\text{W/(N/s)}) \tag{5.3}$$

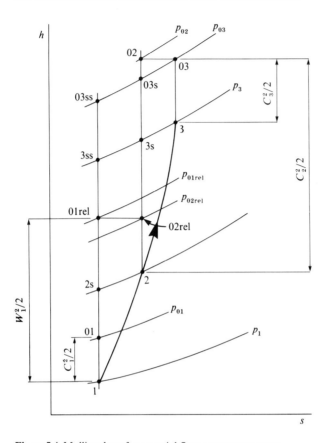

Figure 5.4 Mollier chart for an axial flow compressor stage

Equation (5.2) or (5.3) may be used depending upon the information available.

The flow through the stage is shown thermodynamically on the Mollier chart in Fig. 5.4 and is similar to that for a centrifugal compressor. The enthalpy change is continuous taking account of irreversibilities in the rotor and stator.

Assuming adiabatic flow through the stage, $h_{03} = h_{02}$, and therefore Eq. (5.1) may be written as

$$W/m = h_{02} - h_{01} \quad \text{(W/(kg/s))} \tag{5.4}$$

Writing

$$h_0 = h + C^2/2 = h + (C_a^2 + C_x^2)/2$$

then

$$h_{02} - h_{01} = (h_2 - h_1) + (C_{x2}^2 - C_{x1}^2)/2 = U(C_{x2} - C_{x1})$$

or

$$(h_2 - h_1) - (C_{x2} - C_{x1})[2U - (C_{x2} + C_{x1})]/2 = 0$$

Rearranging,

$$(h_2 - h_1) - (C_{x2} - C_{x1})[(U - C_{x2}) + (U - C_{x1})]/2 = 0$$
$$(h_2 - h_1) + (W_{x2} - W_{x1})(W_{x2} + W_{x1})/2 = 0$$
$$(h_2 - h_1) + (W_{x2}^2 - W_{x1}^2)/2 = 0$$

But $(W_{x2}^2 - W_{x1}^2) = (W_2^2 - W_1^2)$ since C_a is constant. Therefore

$$h_2 + W_2^2/2 = h_1 + W_1^2/2 \qquad (5.5)$$

and Eq. (5.5) can be written

$$h_{02rel} = h_{01rel} \qquad (5.6)$$

where the relative total enthalpy is based on the relative velocity.

Equation (5.6) shows that the total enthalpy based on relative velocities in the rotor is constant across the rotor and this result is also valid for the axial flow gas turbine rotor. A comparison of Eq. (5.5) with Eq.(4.7) indicates why the enthalpy change in a single-stage axial flow compressor is so low compared to the centrifugal compressor. The relative velocities may be of the same order of magnitude, but the axial flow compressor receives no contribution from the change in tangential velocity.

The isentropic efficiency is written as

$$\eta_c = \frac{\text{Ideal isentropic work input}}{\text{Actual work input}}$$

$$= (h_{03ss} - h_{01})/(h_{03} - h_{01})$$

which reduces to

$$\eta_c = T_{01}(T_{03ss}/T_{01} - 1)/(T_{03} - T_{01})$$

Putting

$$p_{03}/p_{01} = (T_{03ss}/T_{01})^{\gamma/(\gamma-1)}$$

the pressure ratio becomes

$$p_{03}/p_{01} = [1 + \eta_c(T_{03} - T_{01})/T_{01}]^{\gamma/(\gamma-1)} \qquad (5.7)$$

The energy input to the fluid will be absorbed usefully in raising the pressure and velocity of the air and some will be wasted in overcoming various frictional losses. However, the whole of the work input will appear as a stagnation temperature rise of the air regardless of the isentropic efficiency. Equation (5.4) written in terms of temperatures and air angles from Eq. (5.2) is

$$(T_{03} - T_{01}) = UC_a(\tan \beta_1 - \tan \beta_2)/C_p \qquad (5.8)$$

`In practice C_a is not constant along the length of the blade and, to account for this, a work done factor λ is introduced, defined as

$$\text{Work done factor} = \frac{\text{Actual work absorbing capacity}}{\text{Ideal work absorbing capacity}}$$

Hence

$$(T_{03} - T_{01}) = \lambda U C_a(\tan \beta_1 - \tan \beta_2)/C_p \tag{5.9}$$

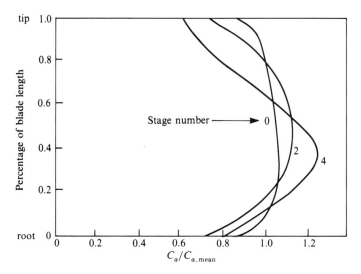

Figure 5.5 Variation of axial velocity along a blade

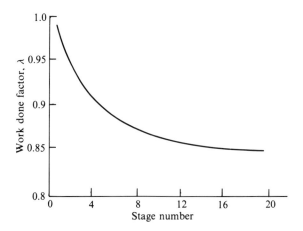

Figure 5.6 Variation of work done factor with number of stages

Figure 5.5 illustrates that it is only at the inlet of the machine that the velocity profile is fairly constant over the blade length. The solid boundaries of the rotor and stator exert more and more influence on the velocity profile as the air moves through the compressor. The variation in work done factor in Fig. 5.6 shows that λ decreases as the number of compressor stages increases.

5.3 REACTION RATIO

The reaction ratio is a measure of the static enthalpy rise that occurs in the rotor expressed as a percentage of the total static enthalpy rise across the stage. It is defined as

$$R = \frac{\text{Static enthalpy rise in rotor}}{\text{Static enthalpy rise in stage}}$$

$$= (h_2 - h_1)/(h_3 - h_1) \tag{5.10}$$

Since $h_{01\text{rel}} = h_{02\text{rel}}$, then

$$(h_2 - h_1) = (W_1^2 - W_2^2)/2$$

Also if $C_1 = C_3$, then

$$(h_3 - h_1) = (h_{03} - h_{01}) = U(C_{x2} - C_{x1})$$

and substituting for $(h_2 - h_1)$ and $(h_3 - h_1)$ in Eq. (5.10) the reaction ratio becomes

$$R = (W_1^2 - W_2^2)/[2U(C_{x2} - C_{x1})]$$

$$= [(C_a^2 + W_{x1}^2) - (C_a^2 + W_{x2}^2)]/[2U(C_{x2} - C_{x1})]$$

$$= (W_{x1} + W_{x2})(W_{x1} - W_{x2})/[2U(C_{x2} - C_{x1})]$$

But $C_{x2} = U - W_{x2}$ and $C_{x1} = U - W_{x1}$. Therefore

$$(C_{x2} - C_{x1}) = (W_{x1} - W_{x2})$$

Hence

$$R = (W_{x1} + W_{x2})/2U$$

$$= C_a(\tan \beta_1 + \tan \beta_2)/2U$$

$$= (C_a/U)(\tan \beta_\infty)$$

$$= \phi \tan \beta_\infty \tag{5.11}$$

Here, $\tan \beta_\infty = (\tan \beta_1 + \tan \beta_2)/2$, while the ratio of axial velocity to blade speed is called the flow coefficient. It may be shown similarly that the reaction ratio can be expressed as

$$R = [1 + \phi(\tan \beta_2 - \tan \alpha_1)]/2 \tag{5.12}$$

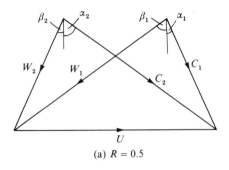

(a) $R = 0.5$

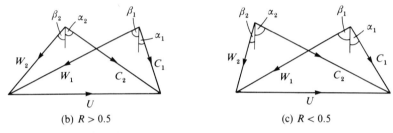

(b) $R > 0.5$ (c) $R < 0.5$

Figure 5.7 Effect of reaction ratio on the velocity triangles

For a reaction ratio of 50 per cent, $(h_2 - h_1) = (h_3 - h_2)$, that is the static enthalpy and temperature increase in the rotor and stator are equal. Also, from Eq. (5.12), $\beta_2 = \alpha_1$, and when the outlet and inlet velocity triangles are superimposed, the resulting diagram is symmetrical. This is shown in Fig. 5.7a. For $R > 0.5$, Fig. 5.7b shows the diagram skewed to the right since $\beta_2 > \alpha_1$, and the static enthalpy rise in the rotor is greater than in the stator. The static pressure rise is also greater in the rotor than the stator. If $R < 0.5$ the diagram is skewed to the left as in Fig. 5.7c, and static enthalpy and pressure rises are greater in the stator than in the rotor. A reaction ratio of 50 per cent is usually chosen so that the adverse pressure gradient over the stage is shared equally by the stator and rotor. This decreases the likelihood of boundary-layer separation in both the stator and rotor blades and is the condition for maximum temperature rise and efficiency.

5.4 STAGE LOADING

If the power input is divided by the term mU^2, a dimensionless coefficient ψ, called the stage loading factor, results:

$$\psi = W/mU^2 = (h_{03} - h_{01})/U^2 \tag{5.13}$$
$$= \lambda(C_{x2} - C_{x1})/U$$
$$= \lambda(C_a/U)(\tan \alpha_2 - \tan \alpha_1)$$
$$\psi = \lambda\phi(\tan \alpha_2 - \tan \alpha_1) \tag{5.14}$$

5.5 LIFT AND DRAG COEFFICIENTS

Consider the rotor blade shown in Fig. 5.8 with relative velocity vectors W_1 and W_2 at angles β_1 and β_2. This system is similar to flow over an aerofoil so that lift and drag forces will be set up on the blade while the forces on the air will act in the opposite direction as shown in Fig. 5.9.

The drag force is defined as acting in the line of the mean velocity vector W_∞ at angle β_∞ to the axial direction as defined by Eq. (5.11), and the lift force acts perpendicular to this.

The resultant force experienced by the air is therefore given by the vector R in Fig. 5.9, so that the force acting in the direction of blade rotation (the x direction) is written as

$$F_x = L \cos \beta_\infty + D \sin \beta_\infty$$
$$= L \cos \beta_\infty [1 + (C_D/C_L)\tan \beta_\infty]$$

But the lift coefficient is defined as

$$C_L = L/0.5\rho W_\infty^2 A$$

where the blade area is the product of the chord c and the span l, and putting

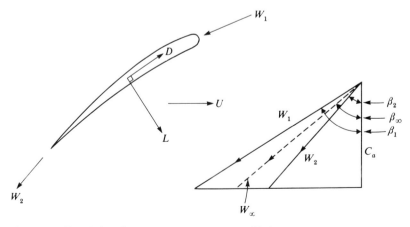

Figure 5.8 Lift and drag forces on a compressor rotor blade

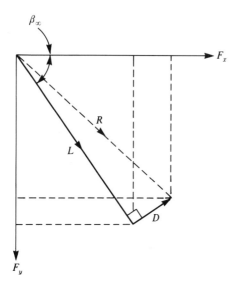

Figure 5.9 Resolving blade forces into the direction of rotation

$W_\infty = C_a/\cos \beta_\infty$ then

$$F_x = \rho C_a^2 c l C_L \sec \beta_\infty [1 + (C_D/C_L)\tan \beta_\infty]/2$$

The power delivered to the air is given by

$$U F_x = m(h_{03} - h_{01}) \quad \text{(W)}$$
$$= \rho C_a l s(h_{03} - h_{01})$$

where the flow through one blade passage of width s has been considered. Therefore

$$\psi = (h_{03} - h_{01})/U^2$$
$$= F_x/\rho C_a l s U$$
$$= (C_a/U)(c/s) \sec \beta_\infty (C_L + C_D \tan \beta_\infty)/2$$
$$= \phi(c/s) \sec \beta_\infty (C_L + C_D \tan \beta_\infty)/2 \tag{5.15}$$

For maximum efficiency, the mean flow angle β_∞ is usually about 45° and substituting for this into Eq. (5.15), the optimum blade loading factor ψ_{opt} becomes

$$\psi_{opt} = (\phi/\sqrt{2})(c/s)(C_L + C_D) \tag{5.16}$$

If C_D is much smaller than C_L, which is usually the case for a well-designed blade, then

$$\psi_{opt} = (\phi/\sqrt{2})(c/s)C_L \tag{5.17}$$

5.6 BLADE CASCADES

The previous sections have concentrated on relating the required energy transfer or stage work to the blade inlet and outlet angles for both the rotor and stator. The next requirement is to decide on the blade shape that will give the required stage work at the maximum efficiency along with the minimum of pressure loss.

In Sec. 2.7.1 use was made of blade element theory to relate the blade lift coefficient to the energy transfer across the impeller of an axial flow pump (Eq. (2.61)), the blades of axial flow pumps and hydraulic turbines being of low solidity. Axial flow compressor (and gas turbine) blading is of high solidity, with the result that the gas flow around a blade is affected by the flow around adjacent blades. In order to obtain information on the effect of different blade designs on air flow angles, pressure losses and expected energy transfer across blade rows, one must resort to cascade wind tunnels and cascade theory.

A cascade is a row of geometrically similar blades arranged at equal distances from each other and aligned to the flow direction as shown in Fig. 5.10. The row of blades is installed on a turntable at the end of a wind tunnel channel such that the angle of incidence of the blades with respect to the approaching air may be varied. Vertical traverses between successive blades may then be made with pitot tubes, and yaw meters to determine pressure losses and air flow angles. Figure 5.10 is known as a linear cascade and can be imagined as a row of compressor blades unwound from the rotor to form the

Figure 5.10 Blade cascade

cascade. The number of blades comprising the cascade has to be sufficient to eliminate any wind tunnel wall boundary-layer effects, and suction slots are often let into the tunnel walls to control the boundary layer.

The data obtained from cascade testing have to be corrected before application to a prototype compressor can be made. The reasons for the corrections are because of the differences between flow in the actual machine and flow through the cascade. These differences are as follows:

1. In the machine, annulus wall boundary layers exist at the blade hub and tip.
2. In the machine, alternate blade rows interfere with the perceived cascade data flow pattern.
3. In the machine, the solidity decreases from hub to tip.
4. Blade velocity varies from hub to tip, thereby affecting the blade inlet angle.

From 3 and 4 it is evident that a cascade test only applies for one radius and inlet angle, and therefore it may be necessary to carry out a number of tests to obtain a reliable picture of the flow in the blades.

5.6.1 Cascade Nomenclature and Curves

Before venturing further into cascade testing and blade design, it is necessary to define various important angles relevant to the design. In Fig. 5.11 a cambered blade is shown with a curved camber line through the centre.

The tangents to the camber line at inlet and outlet are the camber angles α_1' and α_2' to the axial direction respectively. The blade camber angle θ is defined as

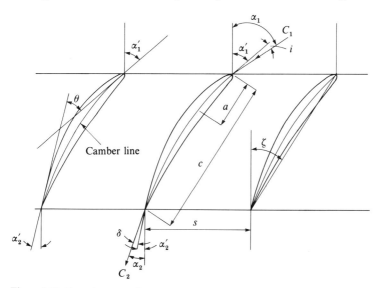

Figure 5.11 Cascade nomenclature

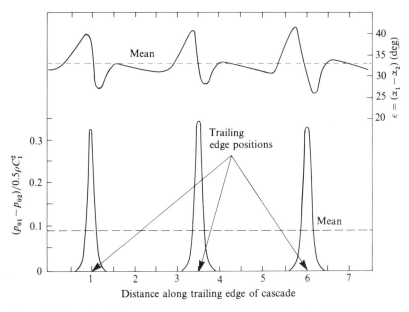

Figure 5.12 Cascade deflection and pressure loss curves at one angle of incidence

$(\alpha'_1 - \alpha'_2)$. The chord c is the distance between blade inlet and trailing edges. The stagger angle ζ is that between the axial direction and the chord and represents the angle at which the blade is set in the cascade. The blade spacing is s and represents the blade pitch. If the air enters with velocity C_1 at angle α_1, the angle of incidence i is $(\alpha_1 - \alpha'_1)$. The air leaves the blade with velocity C_2 at angle α_2 and the difference $(\alpha_2 - \alpha'_2)$ is the deviation angle δ. The air deflection angle is $\varepsilon\,(=\alpha_1 - \alpha_2)$, and it is this angle along with the cascade inlet and outlet stagnation pressures p_{01} and p_{02} that are measured in the traverse along s. The results of the traverses are usually presented as in Fig. 5.12, the stagnation pressure loss being plotted as a dimensionless number given by

$$\text{Stagnation pressure loss coefficient} = (p_{01} - p_{02})/0.5\rho C_1^2 \qquad (5.18)$$

A number of curves such as Fig. 5.12 are obtained for different incidence angles and the mean deflection and pressure loss coefficient for each curve, $\bar{\varepsilon}$ and $(p_{01} - p_{02})/0.5\rho C_1^2$, are plotted against incidence angle as in Fig. 5.13. The deflection increases with angle of incidence up to a maximum ε_s. This is the stall point where separation occurs on the suction surface of the blade, and since this angle may not be well defined in some designs, it is taken as the angle of incidence where the mean pressure loss is twice the minimum. It is evident that for a wide range of incidence, the pressure loss is fairly constant and it is possible to select an angle of deflection ε^* that is also compatible with low pressure loss as representative for the particular design, and by convention

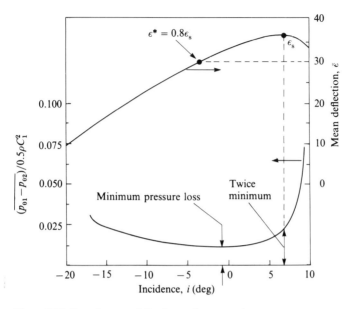

Figure 5.13 Cascade mean deflection and pressure loss curves

$\varepsilon^* = 0.8\varepsilon_s$. The nominal deflection angle ε^* is dependent mainly on the pitch/chord ratio (s/c) and α_2, this being determined from a large number of cascade tests. It is thus possible to plot a set of master curves of ε^* versus α_2 for different values of (s/c). These curves are particularly useful to the designer when any two of the three variables are fixed. For example, if the rotor inlet and outlet angles β_1 and β_2 are known, ε^* can be found, and at angle β_2, (s/c) can be read from Fig. 5.14.

The deviation angle δ is caused by the air not remaining attached to the blade over its total curvature. δ is given by the empirical relationship

$$\delta = m\theta(s/c)^{1/2} \tag{5.19}$$

where

$$m = 0.23(2a/c)^2 + 0.1(\alpha_2/50) \tag{5.20}$$

and a is the distance along the chord to the point of maximum camber. For a circular arc camber line, $(2a/c) = 1$, and this blade form is often chosen.

5.6.2 Cascade Lift and Drag Coefficients

The pressure ratio is governed by the efficiency of the stage and this efficiency depends on the total drag for both the rotor and stator rows. From the measured cascade test results of $(p_{01} - p_{02})$, the lift coefficient C_L and drag

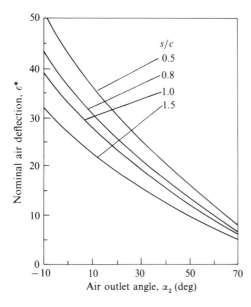

Figure 5.14 Cascade nominal deflection versus air outlet angle

coefficient C_D may be obtained, where C_D is the blade profile drag. The method of approach is to equate all forces acting on the air to the rate of change of momentum of the air.

In Fig. 5.15 two blades of a cascade having chord c and pitch s are shown. At sections 1 and 2 the total air pressures are p_{01} and p_{02} respectively with corresponding velocities of C_1 and C_2, while the density change across the cascade is negligible. The static pressure change across the cascade is therefore

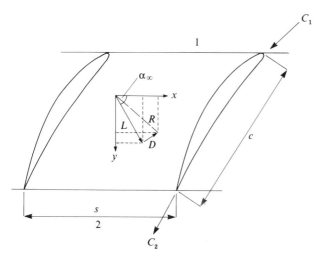

Figure 5.15 Forces and velocities in a cascade

given by

$$(p_2 - p_1) = \rho(C_1^2 - C_2^2)/2 - \overline{(p_{01} - p_{02})} \qquad (5.21)$$

where it should be noted that $p_{01} > p_{02}$ because no work is done in the cascade and the flow is irreversible. Equation (5.21) will be written as

$$\Delta p = \rho(C_1^2 - C_2^2)/2 - \bar{p}_0 \qquad (5.22)$$

where $\Delta p = (p_2 - p_1)$ and $\bar{p}_0 = (p_{01} - p_{02})$.

The summation of all forces acting on the air in the control volume in the x and y directions must equal the rate of change of momentum of the air in these directions. Considering first the y direction, since C_a is constant, there is no velocity change from 1 to 2 in the y direction and consequently no momentum change. Hence for a unit length of blade,

$$L \sin \alpha_\infty - D \cos \alpha_\infty - s\Delta p = 0$$

Therefore

$$D = L \tan \alpha_\infty - s\Delta p/\cos \alpha_\infty \qquad (5.23)$$

In the x direction the velocity changes from C_{x1} to C_{x2}, and noting that these are in the negative x direction,

$$L \cos \alpha_\infty + D \sin \alpha_\infty = - \rho C_a s(C_{x2} - C_{x1})$$

$$= \rho C_a^2 s(\tan \alpha_1 - \tan \alpha_2)$$

and rearranging

$$L = (\rho C_a^2 s/\cos \alpha_\infty)(\tan \alpha_1 - \tan \alpha_2) - D \tan \alpha_\infty \qquad (5.24)$$

Substituting for L and Δp in Eq. (5.23)

$$D = [(\rho C_a^2 s/\cos \alpha_\infty)(\tan \alpha_1 - \tan \alpha_2) - D \tan \alpha_\infty] \tan \alpha_\infty$$
$$- (s/\cos \alpha_\infty)[\rho(C_1^2 - C_2^2)/2 - \bar{p}_0]$$

and

$$D(1 + \tan^2 \alpha_\infty) = (\rho C_a^2 s/\cos \alpha_\infty)(\tan \alpha_1 - \tan \alpha_2) \tan \alpha_\infty$$
$$- (s/\cos \alpha_\infty)[\rho(C_1^2 - C_2^2)/2 - \bar{p}_0]$$

Now $(1 + \tan^2 \alpha_\infty) = \sec^2 \alpha_\infty$ and $\tan \alpha_\infty = (\tan \alpha_1 + \tan \alpha_2)/2$. Therefore

$$D/\cos^2 \alpha_\infty = (\rho C_a^2 s/2 \cos \alpha_\infty)(\tan \alpha_1 - \tan \alpha_2)(\tan \alpha_1 + \tan \alpha_2)$$
$$- (s/\cos \alpha_\infty)[\rho(C_1^2 - C_2^2)/2 - \bar{p}_0]$$

and noting that

$$(C_1^2 - C_2^2) = C_a^2(\tan^2 \alpha_1 - \tan^2 \alpha_2)$$

then

$$D/\cos^2 \alpha_\infty = (\rho C_a^2 s/2 \cos \alpha_\infty)(\tan^2 \alpha_1 - \tan^2 \alpha_2)$$
$$- (\rho C_a^2 s/2 \cos \alpha_\infty)(\tan^2 \alpha_1 - \tan^2 \alpha_2)$$
$$+ (s\bar{p}_0/\cos \alpha_\infty)$$

The first two terms on the RHS are equal and therefore disappear to leave

$$D = s\bar{p}_0 \cos \alpha_\infty$$

Dividing the drag by $0.5 \rho C_\infty^2 c$ gives the drag coefficient

$$C_D = 2(s/c)(\bar{p}_0/\rho C_\infty^2) \cos \alpha_\infty \qquad (5.25)$$

But $C_\infty = C_a/\cos \alpha_\infty$ and $C_a = C_1 \cos \alpha_1$; thus substituting in Eq. (5.25) gives

$$C_D = 2(s/c)(\bar{p}_0/\rho C_1^2)(\cos^3 \alpha_\infty/\cos^2 \alpha_1) \qquad (5.26)$$

A similar procedure may be followed for C_L by substituting for D and Δp in Eq. (5.24) to give

$$L = (\rho C_a^2 s/\cos \alpha_\infty)(\tan \alpha_1 - \tan \alpha_2)$$
$$- \{L \tan \alpha_\infty - (s/\cos \alpha_\infty)[\rho(C_1^2 - C_2^2)/2 - \bar{p}_0]\} \tan \alpha_\infty$$

and

$$L(1 + \tan^2 \alpha_\infty) = (\rho C_a^2 s/\cos \alpha_\infty)(\tan \alpha_1 - \tan \alpha_2)$$
$$+ (s\rho C_a^2/2 \cos \alpha_\infty)(\tan^2 \alpha_1 - \tan^2 \alpha_2)\tan \alpha_\infty$$
$$- (s\bar{p}_0/\cos \alpha_\infty)\tan \alpha_\infty$$
$$L = (\rho C_a^2 s \cos \alpha_\infty)[(\tan \alpha_1 - \tan \alpha_2)$$
$$+ (\tan^2 \alpha_1 - \tan^2 \alpha_2)(\tan \alpha_1 + \tan \alpha_2)/4]$$
$$- (s\bar{p}_0 \cos \alpha_\infty)\tan \alpha_\infty$$
$$= (\rho C_a^2 s \cos \alpha_\infty)(\tan \alpha_1 - \tan \alpha_2)[1 + (\tan \alpha_1 + \tan \alpha_2)^2/4]$$
$$- (s\bar{p}_0 \cos \alpha_\infty)\tan \alpha_\infty$$
$$= (\rho C_a^2 s \cos \alpha_\infty)(\tan \alpha_1 - \tan \alpha_2)/\cos^2 \alpha_\infty - (s\bar{p}_0 \cos \alpha_\infty)\tan \alpha_\infty$$

Now

$$C_L = L/0.5\rho C_\infty^2 c = 2L \cos^2 \alpha_\infty/\rho C_a^2 c = 2 L \cos^2 \alpha_\infty/\rho C_1^2 c \cos^2 \alpha_1$$

Thus

$$C_L = 2(s/c)\cos \alpha_\infty(\tan \alpha_1 - \tan \alpha_2)$$
$$- 2(s/c)(\bar{p}_0/\rho C_1^2)(\cos^3 \alpha_\infty/\cos^2 \alpha_1)\tan \alpha_\infty$$
$$= 2(s/c)\cos \alpha_\infty(\tan \alpha_1 - \tan \alpha_2) - C_D \tan \alpha_\infty \qquad (5.27)$$

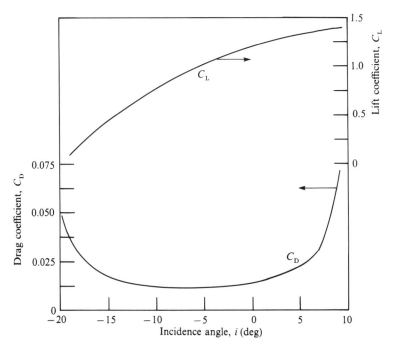

Figure 5.16 Lift and drag coefficients for the cascade from Eqs (5.26) and (5.27)

The air inlet velocity C_1, the incidence angle i and the blade inlet angle α'_1 are known and therefore $\alpha_1 \, (= \alpha'_1 + i)$ is also known. The deviation ε is read from Fig. 5.13 for the angle of incidence, and $\alpha_\infty = \tan^{-1}\left[(\tan\alpha_1 + \tan\alpha_2)/2\right]$ where $\alpha_2 = (\alpha_1 - \varepsilon)$.

Knowing (s/c), values of $\bar{p}_0/0.5\rho C_1^2$ can be read from Fig. 5.13 for various incidence angles and upon substitution of these variables into Eqs (5.26) and (5.27) curves of C_L and C_D may be plotted against the incidence angle as shown in Fig. 5.16. Finally the lift coefficient can be plotted against the air outlet angle α_2 for the nominal value of ε^* for a whole series of different geometry cascades to give the variation of C_L with air outlet angle for a particular (s/c) ratio (Fig. 5.17).

The drag coefficient is usually very small in comparison with C_L and is therefore often ignored so that Eq. (5.27) becomes

$$C_L = 2(s/c)\cos\alpha_\infty(\tan\alpha_1 - \tan\alpha_2) \tag{5.28}$$

To the profile drag as given by Eq. (5.26) two further drags must be added for reasons described in Sec. 5.6. These are the drag effects due to the walls of the compressor, usually called the annulus drag, and the secondary losses caused by trailing vortices at the blade tips. Empirical relationships exist for these drags as follows.

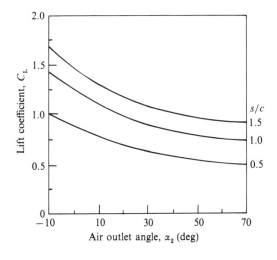

Figure 5.17 Nominal lift coefficients for the cascade

Annulus drag coefficient

$$C_{DA} = 0.02(s/l) \tag{5.29}$$

where l is the blade height.

Secondary losses

$$C_{DS} = 0.018 C_L^2 \tag{5.30}$$

The total drag coefficient is given by

$$C_{DT} = C_D + C_{DA} + C_{DS} \tag{5.31}$$

5.7 BLADE EFFICIENCY AND STAGE EFFICIENCY

It has been shown in the previous section that linear cascade data may be effectively used to determine the lift and drag coefficients for the cascade and then be modified by the addition of annulus drag and secondary losses in order to approximate the drag coefficient for an annular cascade. After determining C_{DT} from Eq. (5.31), the pressure loss coefficient $\bar{p}_0/0.5\rho C_1^2$ is determined from Eq. (5.26). The blade row efficiency η_b is defined as

$$\eta_b = \frac{\text{Actual pressure rise in compressor blade row}}{\text{Theoretical pressure rise in blade row}} \tag{5.32}$$

Now the theoretical pressure rise will occur when $\bar{p}_0$ is zero, that is when there is no total pressure loss across the cascade. Therefore from Eq. (5.22)

$$\Delta p_{\text{theor}} = \rho C_a^2 (\tan^2 \alpha_1 - \tan^2 \alpha_2)/2 \tag{5.33}$$

and blade efficiency is defined as

$$\eta_b = (\Delta \bar{p}_{\text{theor}} - p_0)/\Delta p_{\text{theor}}$$
$$= 1 - \bar{p}_0/\Delta p_{\text{theor}} \tag{5.34}$$

where both actual and theoretical static pressure differences are expressed in terms of known angles. For the normal stage, $C_1 = C_3$, and therefore the stage isentropic efficiency can be approximated to

$$\eta_s = (T_{3ss} - T_1)/(T_3 - T_1) \tag{5.35}$$

It can be shown that for reaction ratios of 50 per cent, when $(T_2 - T_1)$ in the rotor will equal $(T_3 - T_2)$ in the stator, the blade efficiency and stage isentropic efficiencies are the same. For other reaction ratios, the stage efficiency is given by

$$\eta_s = R\eta_{b,\text{rotor}} + (1 - R)\eta_{b,\text{stator}} \tag{5.36}$$

If, for 50 per cent reaction, total values are used instead of static values, then the total-to-total pressure ratio is approximately equal to the static-to-static pressure ratio.

5.8 THREE-DIMENSIONAL FLOW

So far we have been considering only those flows that are two-dimensional in nature in that only whirl and axial flow velocities exist with no radial velocity component. In axial flow turbomachines with hub/tip ratios greater than 0.8 this is a fairly reasonable assumption concerning the flow in the annulus, but for hub/tip ratios less than 0.8, the assumption of two-dimensional flow is no longer valid. This is seen to be so in the case of aircraft compressors, and at the low-pressure end of gas turbines where, because of the high mass flow requirements and the need for as small a frontal area as possible, the trend is to longer blades on a smaller hub. Radial velocities set up in the blade row can now result in an appreciable redistribution of the mass flow, which can seriously affect the blade outlet velocity distribution. Consequently any blade

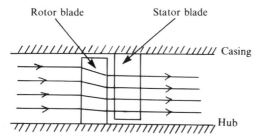

Figure 5.18 Streamline deviation in rotor row for radial equilibrium

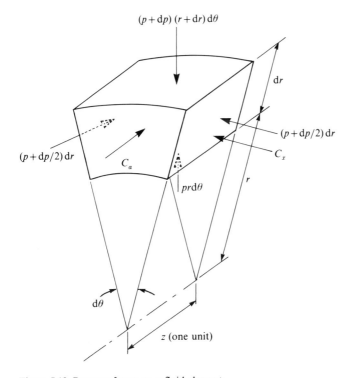

Figure 5.19 Pressure forces on a fluid element

designed on the two-dimensional principle could be seriously in error as regards blade angles.

Radial flow is caused by a temporary imbalance between the centrifugal and radial pressure forces acting on the fluid. When these forces balance each other, there is no radial velocity and the fluid is said to be in radial equilibrium. The method of analysis is known as the 'radial equilibrium method' and assumes that all radial movement of the streamlines takes place in the rotating blade passages, while on either side the streamlines are two-dimensional as illustrated in Fig. 5.18. The analysis examines the pressure forces acting on an element of fluid, shown in Fig. 5.19, and equates these forces to the centrifugal forces acting on the fluid. The stagnation enthalpy h_0 at any radius is then introduced and the following equation is derived:

$$dh_0/dr = C_x^2/r + C_x \, dC_x/dr + C_a \, dC_a/dr \qquad (5.37)$$

Equation (5.37) is the radial equilibrium equation and may be applied to problems in different ways. For instance, if it is assumed that the energy transfer is constant at all radii, then $dh_0/dr = 0$ and

$$C_x^2/r + C_x \, dC_x/dr + C_a \, dC_a/dr = 0 \qquad (5.38)$$

If we now assume (as we have previously) that in the axial flow compressor

C_a is constant over the annulus, then $dC_a/dr = 0$ and hence

$$dC_x/dr = -C_x/r \qquad (5.39)$$

which upon integration gives

$$C_x r = \text{const} \qquad (5.40)$$

and this is the requirement for flow in a free vortex with the whirl velocity being inversely proportional to the radius. The outlet blade angles would therefore be calculated using the free vortex velocity distribution.

It is not necessary to stipulate that dh_0/dr be zero or indeed that C_a is constant at all radii. It might be desirable to specify a radial distribution of energy transfer, choose a radial distribution for C_a and calculate C_x as a function of r which satisfies Eq. (5.37). Having discussed the principle of three-dimensional analysis, however, it should be stressed that it is not a panacea for all design problems since loss of performance due to boundary-layer growth and secondary losses detract even from the improved accuracy of radial equilibrium analysis.

5.9 MULTI-STAGE PERFORMANCE

The total pressure ratio across a single stage is dependent upon the total temperature rise across the stage, and the unthinking might assume that, to find the total pressure ratio across N stages, it would only be necessary to raise the pressure ratio to the power of the number of stages, such that if p_{0r} is the pressure ratio for one stage, then the total pressure ratio is given by $(p_{0r})^N$. However, this is not correct, since for the same temperature rise per stage, as the entropy increases, the pressure rise decreases, as examination of the Mollier chart will show. It is here that the small stage or polytropic efficiency η_p is employed.

In Fig. 5.20, if we compress in a single compression from 1 to 5, the isentropic work done is

$$W/m = (h_{11s} - h_1)$$

and the isentropic efficiency of the compression is

$$\eta_c = (h_{11s} - h_1)/(h_{11} - h_1)$$

If we now compress from 1 to 5 in a number of small finite stages, the isentropic work done is

$$W_s/m = (h_{2s} - h_1) + (h_{3s} - h_2) + (h_{4s} - h_3) + (h_{5s} - h_4)$$

and if for similarly designed stages the efficiency η_s is the same, then

$$\eta_s = (W_s/m)/(h_{11} - h_1)$$

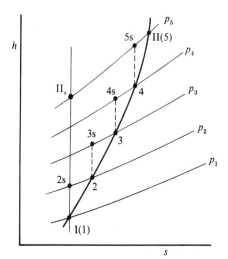

Figure 5.20 Multi-stage compression

where the numerator consists of a number of isentropic enthalpy increases. But as the entropy increases through the compression, so the constant pressure lines diverge and

$$(h_{2s} - h_1) + (h_{3s} - h_2) + \cdots > (h_{IIs} - h_1)$$

and thus

$$\eta_s > \eta_c$$

That is, the overall single isentropic compression efficiency is less than the stage efficiency. The difference also increases with pressure ratio and with the number of stages.

To account for compression in stages, a small stage or polytropic efficiency is defined for an elemental compression process. This efficiency is considered to be constant throughout the whole compression. Assuming constant specific heat, then for the elemental compression in Fig. 5.21

$$\eta_p = dT'/dT \tag{5.41}$$

For an isentropic process,

$$T/p^{(\gamma - 1)/\gamma} = \text{const} \tag{5.42}$$

Thus

$$dT' = \text{const}[p^{-(1/\gamma)}(\gamma - 1)/\gamma]dp$$

Substituting for dT' from Eq. (5.41), and for the constant in Eq. (5.42), then

$$\eta_p(dT/T) = [(\gamma - 1)/\gamma](dp/p)$$

Integrating between the limits of the full compression from I to II,

$$\ln(T_{II}/T_I) = [(\gamma - 1)/\gamma\eta_p]\ln(p_{II}/p_I) \tag{5.43}$$

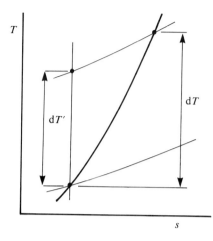

Figure 5.21 Polytropic or small stage compression

and rearranging,

$$p_{II}/p_I = (T_{II}/T_I)^{\eta_p \gamma/(\gamma - 1)} \qquad (5.44)$$

A typical value for polytropic efficiency is 0.88 and in initial design calculations it is often assumed that $\eta_p = \eta_s$.

If it is assumed that we have equal total temperature rises in each stage, and denoting the inlet conditions by 0I and outlet conditions at the last stage as 0II, then for N stages,

$$p_{0II}/p_{0I} = (T_{0II}/T_{0I})^{\eta_p \gamma/(\gamma - 1)} \qquad (5.45)$$

Also

$$T_{0II}/T_{0I} = (T_{0I} + N\Delta T_0)/T_{0I} \qquad (5.46)$$

where ΔT_0 is the stage total temperature rise. It is also usual to assume that the polytropic and total-to-total stage isentropic efficiencies are equal at a value of about 0.88.

While it is possible to make a very rapid calculation of pressure rise through the compressor by this method, the step-by-step calculation of conditions throughout the machine should still be made, particularly if blade forms change.

5.10 AXIAL FLOW COMPRESSOR CHARACTERISTICS

A multi-stage axial compressor characteristic is shown in Fig. 5.22. In comparing this with Fig. 4.10, it is seen that the pressure ratio of the centrifugal compressor is less sensitive to mass flow variations at a given speed than is the axial compressor.

In Fig. 5.22, the design mass flow and pressure ratio are at point 1, and

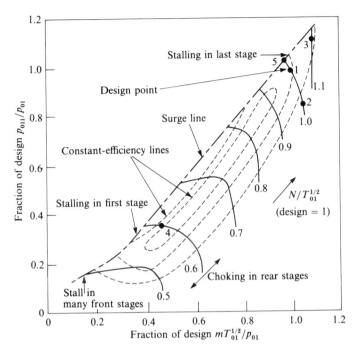

Figure 5.22 Axial flow compressor characteristic

when operating at this condition, all air flow angles and velocities throughout the compressor are at their correct design values. We will examine what happens within the compressor when the mass flow at the design speed is changed, and when the speed itself is changed. In practice the design point is very close to the surge line (point 5) and if the mass flow is only slightly reduced, the pressure ratio and density in the rear stages will both increase. Since $C_a = m/\rho A$, the axial velocity will decrease and hence the incidence angle α_1 will increase sharply in the rear stages, thereby causing stalling in those stages.

Conversely, a small increase in mass flow will lead to a sharp drop in pressure ratio to point 2. The density also drops sharply so that C_a increases. This results in the large decrease of the incidence angle in the rear stages, thereby causing stalling in the rear stages with negative incidence.

If the speed of the compressor is reduced so that the operating point moves to 4, the mass flow and therefore C_a fall faster than the blade speed U, resulting in an increased incidence angle. A further slight reduction in mass flow along the constant-speed characteristic will have little effect on the first-stage density but will cause a further increase in α_1 and possible stalling in the first stage. If the mass flow rate is increased at low speed, the likelihood of first-stage stalling recedes but the density in the rear stages is very low and consequently C_a

increases until sonic conditions and choking of the flow in the rear stages occurs.

When the design speed is exceeded (point 3), the characteristic eventually becomes almost vertical. The increased speed allows more air to be passed at a higher density and pressure ratio. But at the inlet, the mass flow increases faster than the density and choking of the inlet is usually the first to occur. All of the limiting conditions discussed above lead to unstable or inefficient operation and should be avoided at all times.

EXERCISES

Unless otherwise stated, the work done factor is unity and inlet stagnation conditions are 101.3 kPa and 288 K. For air $C_p = 1005$ J/kg K; $R = 287$ J/kg K, $\gamma = 1.4$.

5.1 Using the notation given in the text, show that in an axial flow compressor stage

$$h_2 - h_{2s} = (p_{01rel} - p_{02rel})/\rho$$

5.2 An axial flow compressor stage with 50 per cent reaction has the following data:

Air inlet stagnation temperature	290 K
Relative flow angle at rotor outlet measured from the axial direction	32°
Flow coefficient	0.55
Relative inlet Mach number onto the rotor	0.75

If the stage is normal, what is the stagnation temperature rise in the first stage of the compressor?

5.3 An axial flow compressor stage is to be designed for a stagnation temperature rise of 20 K. The work done factor is 0.92 and the blade velocities at the root, mean radius and tip are 157.5, 210 and 262.5 m/s respectively. The axial velocity is constant from root to tip and is 157.5 m/s. If the reaction ratio at the mean radius is 0.5, what are the inlet and outlet air and blade angles at the root, mean radius and tip for a free vortex design? Calculate also the reaction at the root and tip.

5.4 An alternative design proposal to that in exercise 5.3 is to have 50 per cent reaction along the whole blade. What, then, will the air and blade angles be?

5.5 The design of the first stage of an axial flow compressor calls for the following design data:

Stage stagnation temperature rise	22 K
Mass flow of air	25 kg/s
Rotational speed	150 rev/s
Axial velocity through stage	157 m/s
Work done factor	0.95
Mean blade speed	200 m/s
Reaction at the mean radius	50 per cent
Rotor blade aspect ratio	3
Inlet stagnation temperature	288 K
Inlet stagnation pressure	101.3 kPa

Determine:
  (a) the blade and air angles at the mean radius,
  (b) the mean radius,
  (c) the blade height,
  (d) the pitch and chord and
  (e) the number of blades.

5.6 Using the data of exercise 5.5, if a circular arc camber line for the blade is assumed, and also the data of Fig. 5.13, determine:
  (a) the blade camber angle,
  (b) the deviation,
  (c) the blade stagger,
  (d) the total drag coefficient of the blade,
  (e) blade row efficiency and stage efficiency,
  (f) stage static pressure ratio and
  (g) the stage total pressure ratio.
Assume zero incidence and a normal stage $(C_1 = C_3)$.

5.7 An axial flow compressor has constant axial velocity throughout the compressor of 160 m/s, a mean blade speed of 244 m/s and delivers a pressure ratio of 5:1. Each stage is of 50 per cent reaction and the relative outlet air angles are the same (30°) for each stage. If a polytropic efficiency of 88 per cent is assumed, determine the number of stages in the compressor.

5.8 An axial flow compressor delivers a total pressure ratio of 6, the total head pressure and temperature at entry being 0.408 MPa and 300 K respectively and the overall isentropic efficiency being 82 per cent. The degree of reaction is 50 per cent and all stages contribute an equal amount of work. At a particular stage the blade speed at the mean height is 203 m/s and the axial velocity 171 m/s. If the absolute air angle entering the rotor at this stage is 15° and the work done factor is 0.92, determine:
  (a) the rotor air inlet angle,
  (b) the number of stages required,
  (c) the static temperature of the air at entry to the rotor and
  (d) the rotor inlet relative Mach number.

5.9 An axial flow compressor has 10 stages and the following data apply to each stage at the mean diameter:

Blade speed	200 m/s
Reaction	0.5
Polytropic efficiency	0.88
Stage efficiency	0.84
Angle of absolute air velocity at rotor inlet	13°
Angle of absolute air velocity at rotor outlet	45°
Work done factor	0.86
Inlet stagnation pressure	99.3 kPa
Inlet stagnation temperature	15°C

Determine the total pressure ratio of the first stage and the overall static pressure ratio.

5.10 An axial flow compressor under test in a laboratory exhibits a stage loading of 0.4 for a reaction ratio of 0.65 and flow coefficient 0.55. It is decided to reduce the mass flow by 7 per cent while the blade speed is kept constant, and it is assumed under this new condition that the relative flow exit angles for both the rotor and stator remain unchanged. What is the stage loading and reaction at the new condition? Assume the work done factor is 0.9.

SOLUTIONS

Exercise 5.1 From Eq. (5.6), in the rotor of a turbine or compressor the quantity I is constant from inlet and $h_{01\text{rel}} = h_{02\text{rel}}$. Therefore

$$h_1 + \frac{W_1^2}{2} = h_2 + \frac{W_2^2}{2}$$

or

$$h_1 - h_2 = \frac{W_2^2 - W_1^2}{2}$$

Across the stage the fluid may be assumed incompressible. Therefore

$$p_{01\text{rel}} = p_1 + \frac{\rho W_1^2}{2} \quad \text{and} \quad p_{02\text{rel}} = p_2 + \frac{\rho W_2^2}{2}$$

Hence

$$h_1 - h_2 = \left(\frac{p_{02\text{rel}} - p_2}{\rho}\right) - \left(\frac{p_{01\text{rel}} - p_1}{\rho}\right)$$

Now from Fig. 5.4, along an isentrope

$$T \, ds = dh - \frac{1}{\rho} \, dp = 0$$

Therefore

$$dh = \frac{1}{\rho} \, dp$$

Along 1–2s

$$h_{2s} - h_1 - \frac{p_2 - p_1}{\rho} = 0$$

Thus

$$h_2 - h_{2s} = \frac{p_{01\text{rel}} - p_1 - p_{02\text{rel}} + p_2}{\rho} - \frac{p_2 - p_1}{\rho}$$

$$= \frac{p_{01\text{rel}} - p_{02\text{rel}}}{\rho}$$

Exercise 5.2 Since the stage is normal and reaction 50 per cent, the velocity triangle is symmetrical as in Fig. 5.7.

From Eq. (5.4) $W/m = C_p(T_{02} - T_{01})$ across the stage. Now

$$U(C_{x2} - C_{x1}) = U(W_{x1} - W_{x2})$$

Therefore

$$C_p(T_{02} - T_{01}) = \frac{C_a^2}{\phi}(\tan \beta_1 - \tan \beta_2)$$

The problem resolves itself into finding β_1 and C_a.
   From Eq. (5.11)

$$\tan \beta_1 = \frac{2R}{\phi} - \tan \beta_2$$

$$= \frac{2 \times 0.5}{0.55} - \tan \beta_2$$

$$= 1.19$$

$$\beta_1 = 50 \text{ deg}$$

Now

$$M_{r1} = \frac{W_1}{(\gamma R T_1)^{1/2}}$$

from which

$$W_1^2 = \gamma R M_{r1}^2 \left(T_{01} - \frac{C_1^2}{2C_p} \right)$$

But $W_1 = C_a/\cos \beta_1$ and $C_1 = C_a/\cos \alpha_1$, also $\beta_2 = \alpha_1$. Thus

$$C_1 = C_a/\cos 32° = C_a/0.848$$

and

$$W_1 = C_a/\cos 50° = C_a/0.643$$

Substituting for W_1 and C_1

$$C_a^2 = \gamma R M_{r1}^2 \left(T_{01} - \frac{C_a^2}{1.44 C_p} \right) \cos^2 50°$$

$$= 1.4 \times 287 \times 0.5625 \left(290 - \frac{C_a^2}{1.44 \times 1005} \right) 0.643^2$$

$$= 2.71 \times 10^4 - 0.064 C_a^2$$

$$C_a^2 = 25\,470$$

$$C_a = 159.6 \text{ m/s}$$

Substituting for the calculated values

$$T_{02} - T_{01} = \frac{159.6^2}{1005 \times 0.55}(\tan 50° - \tan 32°)$$

<u>Stagnation temperature rise = 26.1 K</u>

Exercise 5.3 Solution begins at the *mean radius* r_m and from Eq. (5.9)

$$T_{03} - T_{01} = \frac{\lambda U_m C_a}{C_p}(\tan \beta_1 - \tan \beta_2)$$

whence

$$\tan \beta_1 - \tan \beta_2 = \frac{20 \times 1005}{0.92 \times 210 \times 157.5}$$

$$= 0.66$$

Also from Eq. (5.11)

$$R = \frac{C_a}{2U_m}(\tan \beta_1 + \tan \beta_2)$$

Thus

$$\tan \beta_1 + \tan \beta_2 = \frac{0.5 \times 2 \times 210}{157.5}$$

$$= 1.333$$

Eliminating $\tan \beta_2$ gives

$$2\tan \beta_1 = 1.99$$
$$\beta_1 = 44.8°$$

Then

$$\tan \beta_2 = 1.333 - 0.993$$
$$\beta_2 = 18.78°$$

$\alpha_1 = \beta_2$ and $\alpha_2 = \beta_1$ since $R = 0.5$ at the mean radius.
At the *tip*, from the velocity triangle of Fig. 5.3 at inlet

$$\tan \alpha_1 + \tan \beta_1 = U/C_a$$

$$= \frac{262.5}{157.5}$$

$$= 1.666$$

Similarly at outlet

$$\tan \alpha_2 + \tan \beta_2 = 1.66$$

For free vortex design $rC_{x1} = $ const at all radii and therefore between the mean radius and the tip

$$r_m C_{xm} = r_t C_{xt}$$
$$r_m \tan \alpha_{1m} = r_t \tan \alpha_{1t}$$

Since C_a is a constant

$$\tan \alpha_{1t} = \frac{r_m}{r_t}\tan \alpha_{1m}$$

But

$$\frac{r_m}{r_t} = \frac{U_m}{U_t} = 0.8$$

Then

$$\tan \alpha_{1t} = 0.8 \tan 18.78°$$

At tip, inlet air angle

$$\underline{\alpha_{1t} = 15.21°}$$

$$\tan \beta_{1t} = \frac{262.5}{157.5} - \tan 15.21°$$

At tip, blade air angle

$$\underline{\beta_{1t} = 54.37°}$$

At outlet

$$rC_{x2} = \text{const} \qquad \text{at all radii}$$

Therefore

$$\tan \alpha_{2t} = 0.8 \tan \alpha_{2m}$$
$$= 0.8 \tan 44.8°$$
$$= 0.794$$

At tip, outlet air angle

$$\underline{\alpha_{2t} = 38.45°}$$

$$\tan \beta_{2t} = 1.66 - \tan \alpha_{2t}$$
$$= 1.66 - 0.794$$

At tip, outlet blade angle

$$\underline{\beta_{2t} = 40.9°}$$

Moving now to the *root*, the same equations apply. At inlet

$$\tan \alpha_1 + \tan \beta_1 = 1$$

and

$$rC_{x1} = \text{const}$$

$$\tan \alpha_{1r} = \frac{U_m}{U_r} \tan \alpha_{1m}$$

$$= \frac{210}{157.5} \tan 18.78°$$

$$= 0.453$$

At inlet, air angle

$$\alpha_{1r} = 24.38°$$

$$\tan \beta_{1r} = 1 - \tan 24.38°$$

At root, inlet blade angle

$$\underline{\beta_{1r} = 28.68°}$$

At root, outlet air angle

$$\tan \alpha_{2r} = \frac{210}{157.5} \tan \alpha_{2m}$$

$$= 1.333 \tan 44.8°$$

$$\underline{\alpha_{2r} = 52.94°}$$

At root, outlet blade angle

$$\tan \beta_{2r} = 1 - \tan \alpha_{2r}$$
$$= 1 - 1.324$$
$$= -0.324$$

and β_{2r} is given by

$$\underline{\beta_{2r} = -17.95°}$$

At the tip, reaction

$$R_t = \frac{\phi}{2}(\tan \beta_{1t} + \tan \beta_{2t})$$

$$= \frac{157.5}{262.5 \times 2}(\tan 54.37° + \tan 40.9°)$$

$$\underline{= 0.68}$$

At the root, reaction

$$R_r = \frac{157.5}{2 \times 157.5}(\tan 28.68° + \tan(-17.95°))$$

$$\underline{= 0.112}$$

Exercise 5.4 Equations (5.11) and (5.9) are used at all radii. The angles at the mean radii will of course remain unchanged.

At the tip using Eq. (5.11)

$$0.5 = \frac{157.5}{2 \times 262.5}(\tan \beta_1 + \tan \beta_2)$$

and using Eq. (5.9)

$$\tan \beta_1 - \tan \beta_2 = \frac{20 \times 1005}{0.92 \times 157.5 \times 262.5}$$

$$= 0.528$$

Solving simultaneously for $\tan \beta_1$

$$2 \tan \beta_1 = 1.666 + 0.528$$

$$\tan \beta_1 = 1.097$$

At the tip, blade angle and air angles are equal:

$$\beta_1 = 47.64° = \alpha_2$$

Blade outlet angle

$$\tan \beta_2 = \tan 47.64 - 0.528$$

$$= 1.097 - 0.528$$

$$\beta_2 = 29.64° = \alpha_1$$

At the root using the same equations

$$\tan \beta_1 + \tan \beta_2 = \frac{0.5 \times 2 \times 157.5}{157.5}$$

$$= 1$$

and

$$\tan \beta_1 - \tan \beta_2 = \frac{20 \times 1005}{0.92 \times 157.5 \times 157.5}$$

$$= 0.881$$

whence

$$2 \tan \beta_1 = 1.881$$

$$\beta_1 = 43.23° = \alpha_2$$

$$\tan \beta_2 = 1 - 0.94$$

$$\beta_2 = 3.43° = \alpha_1$$

Exercise 5.5 (a) Using Eq. (5.9) at the mean radius

$$T_{03} - T_{01} = \frac{\lambda U C_a}{C_p}(\tan \beta_1 - \tan \beta_2)$$

$$22 = \frac{0.95 \times 200 \times 157}{1005}(\tan \beta_1 - \tan \beta_2)$$

$$\tan \beta_1 - \tan \beta_2 = 0.741$$

Using Eq. (5.11)

$$R = \frac{C_a}{2U}(\tan \beta_1 + \tan \beta_2)$$

and hence

$$\tan \beta_1 + \tan \beta_2 = \frac{0.5 \times 2 \times 200}{157}$$

$$= 1.27$$

Solving simultaneously for β_1 and β_2

$$2 \tan \beta_1 = 2.011$$
$$\underline{\beta_1 = 45.16° = \alpha_2}$$

(since diagram is symmetrical) and

$$\tan \beta_2 = 1.27 - \tan 45.16$$
$$\underline{\beta_2 = 14.81° = \alpha_1}$$

(b) Mean radius

$$r_m = \frac{U}{2\pi N}$$

$$= \frac{200}{2\pi \times 150}$$

$$= 0.212 \, \text{m}$$

(c) The blade height is found from the annulus area of flow as used in the continuity equation

$$m = \rho_1 A C_a$$

Now

$$C_1 = \frac{C_a}{\cos \alpha_1}$$

$$= \frac{157}{\cos 14.81}$$

$$= 162.4 \, \text{m/s}$$

$$T_1 = T_{01} - \frac{C_1^2}{2C_p}$$

$$= 288 - \frac{162.4^2}{2 \times 1005}$$

$$= 274.88 \text{ K}$$

Assuming isentropic flow at inlet we may use isentropic relationships at a point

$$\frac{p_1}{p_{01}} = \left(\frac{T_1}{T_{01}}\right)^{\gamma/(\gamma-1)}$$

Static pressure

$$p_1 = 101.3 \left(\frac{274.9}{288}\right)^{3.5}$$

$$= 86.1 \text{ kPa}$$

Then

$$\rho_1 = \frac{p_1}{RT_1}$$

$$= \frac{86.1 \times 10^3}{287 \times 274.88}$$

$$= 1.09 \text{ kg/m}^3$$

From the continuity equation

$$A = \frac{25}{1.09 \times 157}$$

$$= 0.146 \text{ m}^2$$

Blade height

$$h = \frac{A}{2\pi r_m}$$

$$= \frac{0.146}{2\pi \times 0.212}$$

$$= 0.11 \text{ m}$$

The variation of air angle across the annulus is determined as in exercises 5.3 and 5.4 and will not be repeated here. Suffice it to say that they will be depedent upon the type of flow chosen, i.e. free vortex, contant reaction, etc.

(d) The nominal deflection curve (Fig. 5.14) for a number of blade cascades may now be used to find the pitch and chord. At the mean radius, and noting that blades, β, an equivalent to cascades, α,

$$\varepsilon^* = \beta_1 - \beta_2$$

$$= 45.16° - 14.81°$$

$$= 30.35°$$

and hence from Fig. 5.14 at the air outlet angle β_2 of 14.81°, the solidity is found:

$$s/c = 0.8$$

$$\text{Blade aspect ratio} = \frac{\text{Span}}{\text{Chord}}$$

$$c = \frac{0.11}{3}$$

$$\underline{\text{Blade chord} = 0.0366\,\text{m}}$$

Blade pitch

$$s = 0.8 \times 0.0366$$
$$\underline{= 0.0293\,\text{m}}$$

(e)Number of blades $= \dfrac{\text{Circumference at mean radius}}{\text{Pitch at mean radius}}$

$$= \frac{2\pi \times 0.212}{0.0293}$$

$$\underline{= 45.46}$$

Blade numbers are usually chosen to avoid those with common multiples. This is to avoid the chances of resonant frequencies being set up between the fixed and moving blades. In this exercise a choice of 45 is divisible by 5 and 9 and therefore a prime under such as 43 or 47 blades would be chosen. This alters the calculations slightly and so if we choose 47 blades

$$\underline{s = 0.0283\,\text{m}}, \qquad \underline{c = 0.0354\,\text{m}} \qquad \text{and} \qquad \underline{\text{Aspect ratio} = 3.1}$$

Exercise 5.6 (a) The deviation angle $\delta = \alpha_2 - \alpha_2'$ as shown in Fig. 5.11 and is given by Eq. (5.19). In Eq. (5.20) for m, $(2a/c)$ is unity (for circular arc camber) and at the mean radius, α_2 for the cascade equals β_2 for the moving blade, $\alpha_2 = 14.81$. Therefore

$$m = 0.23\left(\frac{2a}{c}\right)^2 + 0.1\left(\frac{\alpha_2}{50}\right)$$

$$= 0.23 \times (1) + 0.1\left(\frac{14.81}{50}\right)$$

$$= 0.26$$

and

$$\delta = m\theta(s/c)^{1/2}$$
$$= 0.26(0.8)^{1/2}\theta$$

$$\underline{\text{Deviation } \delta = 0.233\theta}$$

(b) Camber angle

$$\theta = \alpha_1' - \alpha_2'$$
$$= \alpha_1' - \alpha_2 + \delta$$
$$= \alpha_1' - \alpha_2 + 0.233\theta$$

The cascade air inlet angle α_1 = compressor relative air inlet angle $\beta_1 = 45.16°$, and since there is no incidence on the blade, cascade blade inlet angle α_1' equals air inlet angle α_1.

$$\theta = 45.16 - 14.81 + 0.233\theta$$

$$\text{Camber angle } \theta = 39.57°$$

(c) From the geometry of Fig. 5.11 the blade stagger for a circular arc cascade is given by

$$\zeta = \alpha_1' - \frac{\theta}{2}$$

$$= 45.16 - \frac{39.57}{2}$$

$$\text{Stagger angle } = 25.4°$$

All information is now available for construction of a chosen profile blade on the circular arc camber line at the angles calculated.

(d) The cascade lift coefficient for $(s/c) = 0.8$ is given from Fig. 5.17 as $C_L = 0.9$ at $\alpha_2 = 14.81°$. From Eq. (5.30) the secondary losses are

$$C_{DS} = 0.018 C_L^2$$
$$= 0.018 \times (0.9)^2$$
$$= 0.014\,58$$

and from Eq. (5.29) the annulus losses are given by

$$C_{DA} = 0.02(s/h) \quad (\text{span} = \text{blade height})$$

$$= 0.02 \times \frac{0.0283}{0.11}$$

$$= 0.005\,145$$

Then using Fig. 5.13 at zero incidence, $\bar{p}_0/\frac{1}{2}\rho C_1^2 = 0.012$ and substituting this into Eq. (5.26) gives for the straight cascade

$$C_D = \left(\frac{s}{c}\right)\left(\frac{\bar{p}_0}{\frac{1}{2}\rho C_1^2}\right)\frac{\cos^3 \alpha_\infty}{\cos^2 \alpha_1}$$

$$= 0.8 \times 0.012\left(\frac{\cos^3 \alpha_\infty}{\cos^2 45.16}\right)$$

Now

$$\tan \alpha_\infty = \frac{\tan \alpha_1 + \tan \alpha_2}{2}$$

$$= \frac{\tan 45.16 + \tan 14.81}{2}$$

$$\alpha_\infty = 32.41°$$

Then

$$C_D = \frac{0.8 \times 0.012 \times \cos^3 32.41}{\cos^2 45.16}$$

$$= 0.0116$$

Therefore substituting for the losses into Eq. (5.31) the total drag coefficient is

$$C_{DT} = C_D + C_{DA} + C_{DS}$$
$$= 0.0116 + 0.005\,145 + 0.014\,58$$

Blade drag coefficient $= 0.0313$

(e) Using Eq. (5.26) again to find the pressure loss coefficient in the moving blades since the same equations and arguments will apply to circular cascades as well as straight cascades

$$C_{DT} = \left(\frac{s}{c}\right) \frac{\bar{p}_0}{\frac{1}{2}\rho C_1^2} \frac{\cos^3 \alpha_\infty}{\cos^2 \alpha_1}$$

and

$$\frac{\bar{p}_0}{\frac{1}{2}\rho C_1^2} = \frac{0.0313 \times \cos^2 45.16}{0.8 \cos^3 32.41}$$

$$= 0.0323$$

Now the theoretical pressure rise through the row of blades is given by Eq. (5.33) when $\bar{p}_0$ is zero and this leads to the blade efficiency of Eq. (5.34). Blade efficiency

$$\eta_b = 1 - \frac{\bar{p}_0}{\Delta p_{\text{theor}}}$$

Dividing Eq. (5.33) by $\rho C_1^2/2$ we get

$$\frac{\Delta p_{\text{theor}}}{\frac{1}{2}\rho C_1^2} = \left(\frac{C_a}{C_1}\right)^2 (\tan^2 \alpha_1 - \tan^2 \alpha_2)$$

$$= \cos^2 \alpha_1 (\tan^2 \alpha_1 - \tan^2 \alpha_2)$$

$$= \cos^2 45.16 (\tan^2 45.16 - \tan^2 14.81)$$

$$= 0.468$$

Thus

$$\eta_b = 1 - \frac{0.0323}{0.468}$$

$$= 0.931$$

Blade row efficiency $= 0.931$

The stator blade row efficiency will be almost identical for 50 per cent reaction (at the mean diameter). Differences in the height and blade pitch between rotor and stator rows can affect C_{DA} but, as shown in (d), this drag constitutes only a small part of the total drag and its effect therefore is negligible.

The stage efficiency for 50 per cent reaction is found from Eq. (5.36) to be the same as the blade efficiency,

$$\eta_s = 0.931$$

(f) The stage static pressure ratio is found from Eq. (5.35)

$$\frac{T_{3ss}}{T_1} = \left(1 + \eta_s \frac{T_3 - T_1}{T_1}\right) = \left(\frac{p_3}{p_1}\right)^{(\gamma - 1)/\gamma}$$

Now $C_1 = C_3$ for a normal stage and therefore the stage static temperature rise equals the stage stagnation temperature rise

$$\frac{p_3}{p_1} = \left[1 + 0.931\left(\frac{22}{274.88}\right)\right]^{3.5}$$

$$= 1.29$$

(g)
$$\frac{p_{03}}{p_{02}} = \left[1 + 0.931\left(\frac{22}{288}\right)\right]^{3.5}$$

$$= 1.27$$

Exercise 5.7 Since we have 50 per cent reaction the velocity diagrams are symmetrical as in Fig. 5.7a, and $\alpha_1 = \beta_2$, $\alpha_2 = \beta_1$. The number of stages may be found from Eq. (5.46)

$$\frac{T_{0II}}{T_{0I}} = \frac{T_{0I} + N\Delta T_0}{T_{0I}}$$

Using Eq. (5.8) the stage stagnation temperature rise is

$$\Delta T_0 = T_{03} - T_{01} = \frac{UC_a}{C_p}(\tan \beta_1 - \tan \beta_2) \qquad \text{where } \beta_2 = 30°$$

Now

$$W_{x2} = C_a \tan \beta_2$$
$$= 160 \tan 30°$$
$$= 92.38 \text{ m/s}$$

Therefore

$$W_1 = C_2 = [(U - W_{x2})^2 + C_a^2]^{1/2}$$
$$= [(244 - 92.38)^2 + 160^2]^{1/2}$$
$$= 220.4 \text{ m/s}$$
$$\cos \beta_1 = C_a/W_1$$
$$= \frac{160}{220.4}$$
$$\beta_1 = 43.45°$$

Therefore

$$\Delta T_0 = \frac{244 \times 160}{1005}(\tan 43.45° - \tan 30°)$$
$$= 14.37 \text{ K}$$

Equation (5.44) gives the overall temperature ratio as

$$\frac{T_{0II}}{T_{0I}} = \left(\frac{p_{0II}}{p_{0I}}\right)^{(\gamma-1)/\eta_p\gamma}$$
$$= 5^{0.4/(0.88 \times 1.4)}$$
$$= 5^{0.3247}$$
$$= 1.68$$

Thus

$$N = \frac{(1.68 \times 288) - 288}{14.37}$$
$$= 13.3$$

$$\underline{\text{Number of full stages} = 14}$$

Exercise 5.8 (a) The velocity diagrams of Fig. 5.3 are once again symmetrical and $\alpha_1 = 15° = \beta_2$.

$$C_{x1} = C_a \tan \alpha_1$$
$$= 171 \times \tan 15°$$
$$= 45.8 \text{ m/s}$$

$$\tan \beta_1 = \frac{U - C_{x1}}{C_a}$$
$$= \frac{203 - 45.8}{171}$$

$$\tan \beta_1 = 0.919$$

$$\beta_1 = 42.6° = \alpha_2 \qquad \text{for 50 per cent reaction}$$

$$\underline{\text{Rotor inlet angle } \beta_1 = 42.6°}$$

(b) The overall isentropic efficiency is obtained by working between the compressor inlet and outlet conditions.

$$\text{Overall isentropic efficiency} = \frac{\text{Isentropic work done}}{\text{Actual work done}}$$

$$= \frac{T_{0\text{IIs}} - T_{0\text{I}}}{T_{0\text{II}} - T_{0\text{I}}}$$

$$= \left(\frac{T_{0\text{IIs}}}{T_{0\text{I}}} - 1\right) \Big/ \left(\frac{T_{0\text{II}}}{T_{0\text{I}}} - 1\right)$$

Substituting for $(T_{0\text{II}}/T_{0\text{I}})$ from Eq. (5.45) and $(T_{0\text{IIs}}/T_{0\text{I}})$, using isentropic flow relationships we get

$$\eta_0 = \left[\left(\frac{p_{0\text{II}}}{p_{0\text{I}}}\right)^{(\gamma-1)/\gamma} - 1\right] \Big/ \left[\left(\frac{p_{0\text{II}}}{p_{0\text{I}}}\right)^{(\gamma-1)/\gamma\eta_p} - 1\right]$$

$$0.82 = (6^{0.286} - 1)/(6^{0.286/\eta_p} - 1)$$

$$6^{0.286/\eta_p} = \left(\frac{1.669 - 1}{0.82}\right) + 1$$

$$= 1.816$$

$$\frac{0.286}{\eta_p} \log 6 = \log 1.816$$

Polytropic efficiency

$$\eta_p = 0.859$$

Therefore

$$\frac{T_{0\text{II}}}{T_{0\text{I}}} = \left(\frac{p_{0\text{II}}}{p_{0\text{I}}}\right)^{0.286/0.859}$$

$$= 1.82 \text{ K}$$

Substitute into Eq. (5.46) to find the number of stages

$$\frac{T_{0\text{II}}}{T_{0\text{I}}} = \frac{T_{0\text{I}} + N\Delta T_0}{T_{0\text{I}}}$$

where

$$\Delta T_0 = \text{Stagnation temperature rise in the stage}$$

and

$$C_p \Delta T_0 = \psi \, C_a U (\tan \beta_1 - \tan \beta_2) \qquad \text{from Eq. (5.9)}$$

$$= 0.92 \times 171 \times 203(\tan 42.6° - \tan 15°)$$

$$= 20.81 \text{ kJ/kg}$$

Number of stages

$$N = \left[\frac{(1.82 \times 300) - 300}{20.81 \times 10^3} \right] 1005$$

$$= 11.8$$
$$= 12 \text{ stages}$$

(c) At entry to rotor

$$T_1 = T_{01} - \frac{C_1^2}{2C_p}$$

Now

$$C_1 = \frac{C_a}{\cos \alpha_1}$$

$$= \frac{171}{\cos 15°}$$

$$= 177$$

Therefore

$$T_1 = 300 - \frac{177^2}{2 \times 1005}$$

$$= 300 - 15.58$$
$$= 284.4 \text{ K}$$

(d) The rotor relative inlet Mach number is given by

$$M_{1r} = W_1/a_1$$

where

$$a_1 = (\gamma R T_1)^{1/2}$$
$$= (1.4 \times 287 \times 284.4)^{1/2}$$
$$= 338 \text{ m/s}$$

Also

$$W_1^2 = (U - C_{x1})^2 + C_a^2$$
$$= (203 - 45.8)^2 + 171^2$$
$$= 53\,953$$
$$W_1 = 232.3 \text{ m/s}$$

Therefore substituting

$$M_{r1} = \frac{232.3}{338}$$

$$= 0.687$$

Exercise 5.9 (a) The axial velocity is first found from the velocity triangles and since they are symmetrical for 50 per cent reaction from Fig. 5.7a, $\alpha_1 = \beta_2$ and $\alpha_2 = \beta_1$. Therefore

$$\frac{U - C_a \tan \alpha_1}{C_a} = \tan \beta_1$$

$$C_a = \frac{200}{\tan 45° + \tan 13°}$$

$$= 162.5 \, \text{m/s}$$

Work done in the first stage is given by Eq. (5.9) as

$$C_p(T_{03} - T_{01}) = \psi C_a U(\tan \beta_1 - \tan \beta_2)$$
$$= 0.86 \times 162.5 \times 200(\tan 45° - \tan 13°)$$
$$= 21\,497 \, \text{J/kg}$$

For the first stage, the pressure ratio is given by Eq. (5.7):

$$\frac{p_{03}}{p_{01}} = \left(1 + \eta_c \frac{T_{03} - T_{01}}{T_{01}}\right)^{\gamma/(\gamma-1)}$$

$$= \left(1 + \frac{0.84 \times 21\,497}{288 \times 1005}\right)^{3.5}$$

$$= 1.24$$

To find the overall static pressure ratio, we may use Eq. (5.44):

$$\frac{p_{II}}{p_{I}} = \left(\frac{T_{II}}{T_{I}}\right)^{\eta_p \gamma/(\gamma-1)}$$

Now

$$T_{0II} = T_{0I} + 10 \times \Delta T_0$$

$$= 288 + \frac{10 \times 21\,497}{1005}$$

$$= 502 \, \text{K}$$

But

$$C_1 = \frac{C_a}{\cos \alpha_1}$$

$$= \frac{162.5}{\cos 13°}$$

$$= 166.8 \, \text{m/s}$$

Therefore

$$T_1 = T_{0\text{II}} - \frac{C_1^2}{2C_p}$$

$$= 288 - \frac{166.8^2}{2 \times 1005}$$

$$= 274.1 \text{ K}$$

and

$$T_{\text{II}} = T_{0\text{II}} - \frac{C_1^2}{2C_p}$$

$$= 502 - \frac{166.8^2}{2 \times 1005}$$

$$= 488.1 \text{ K}$$

Thus

$$\frac{p_{\text{II}}}{p_{\text{I}}} = \left(\frac{488.1}{274.1}\right)^{0.88 \times 1.4/0.4}$$

Overall static pressure ratio $= 5.91$

Exercise 5.10 The velocity diagrams at both conditions are sketched below (Fig. 5.23) and it will be noted that they are no longer symmetrical. When the mass flow is reduced, axial velocity C_a is also reduced, and noting that it is angles β_2 and α_1 that remain constant, the dotted lines give the new velocity triangles. The angle α_1 is constant as this is the exit angle of the air from the previous stator.

Equation (5.14) gives the stage loading

$$\psi = \lambda\phi(\tan\alpha_2 - \tan\alpha_1)$$
$$= \lambda\phi(\tan\beta_1 - \tan\beta_2)$$

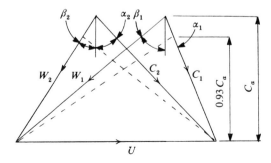

Figure 5.23

Equation (5.11) gives the reaction ratio

$$R = \frac{\phi}{2}(\tan \beta_1 + \tan \beta_2)$$

Solving simultaneously for β_1 and β_2

$$\tan \beta_1 = \left(R + \frac{\psi}{2\lambda} \right) \Big/ \phi$$

$$\tan \beta_2 = \left(R - \frac{\psi}{2\lambda} \right) \Big/ \phi$$

Therefore under the initial flow conditions

$$\tan \beta_1 = \left(0.65 + \frac{0.4}{1.8} \right) \Big/ 0.55$$

$$\beta_1 = 57.76°$$

and

$$\tan \beta_2 = \left(0.65 - \frac{0.4}{1.8} \right) \Big/ 0.55$$

$$\beta_2 = 37.87°$$

From the velocity triangles

$$U = C_a(\tan \beta_2 + \tan \alpha_2)$$

Therefore

$$\tan \alpha_2 = \frac{1}{0.55} - \tan 37.87°$$

$$\alpha_2 = 46.14°$$

Also

$$U = C_a(\tan \alpha_1 + \tan \beta_1)$$

then

$$\tan \alpha_1 = \frac{1}{0.55} - \tan 57.76°$$

$$\alpha_1 = 13.1°$$

At the reduced mass flow

$$\phi = 0.55 \times 0.93$$
$$= 0.5115$$

Then

$$\tan \beta_1 = \frac{1}{0.5115} - \tan 13.1°$$

$$\beta_1 = 59.86°$$

Also

$$\tan \alpha_2 = \frac{1}{0.5115} - \tan 37.87°$$

$$\alpha_2 = 49.65°$$

New stage loading

$$\psi = 0.9 \times 0.5115 (\tan 59.86° - \tan 37.87°)$$

$$= \underline{0.435}$$

New reaction

$$R = \frac{0.5115}{2} (\tan 59.86° + \tan 37.87°)$$

$$= \underline{0.64}$$

AXIAL FLOW STEAM AND GAS TURBINES

6.1 INTRODUCTION

In this section, steam and gas turbines will be considered together, with the assumption that the same theory applies to both types of machines. This commonality is valid providing the steam is in the superheated state and therefore behaves approximately as a perfect gas. Care should be taken, when the condition of the steam falls below the saturation line on the Mollier chart, that the theory and equations developed in later sections are not applied in that case. This would most likely occur at the low-pressure end of the steam turbine.

Axial flow turbines were first built and used successfully by Sir Charles Parsons when he installed a reaction steam turbine in a small marine vessel for propulsion. Since then development of the steam turbine has been rapid and very successful, the power output at the present time ranging from a few kilowatts to 660 MW. The development of the high outputs is due in part to the relatively easy matter of superheating steam in a boiler and superheater, then expanding it through the turbine to below atmospheric pressure in the condenser to extract the maximum energy from the steam.

Development of the axial flow gas turbine was hindered by the need to obtain both a high enough flow rate and compression ratio from a compressor to maintain the air requirement for the combustion process and subsequent expansion of the exhaust gases. Initially the air was provided by centrifugal compressors, and later the axial flow compressor was developed, which, in the case of turbojet and turbofan engines, uses the power developed by the turbine to drive it. Steam turbines are universally used in fossil fuel power stations and for steam-driven propulsion in ships, although gas turbine propulsion units are often fitted in the smaller class of naval vessel. Gas turbines are universally used as the power unit for large jet aircraft propulsion, their advantage being that they have a high power-to-weight ratio.

The principle of energy extraction from the gas is one of gradually reducing the high-pressure energy by converting it into kinetic energy. This is accomplished by passing the gas alternately through rows of fixed and moving blades. The kinetic energy of the gas is reduced in the moving blades, which are attached to the turbine hub, and recovered in the fixed stationary blades attached to the casing. This necessitates a gradual density decrease as the gas moves through the turbine and the blade height therefore increases towards the low-pressure end, if a constant axial flow velocity is to be maintained through the turbine. The stator row is often termed the nozzle row and in certain types of steam turbine the nozzle row consists of a set of converging nozzles spaced around the drum.

Figure 6.1 shows a steam turbine rotor in the bottom half of its casing. It will be noted that the blade height increases in the direction of gas flow. The two important types of axial flow turbines are the reaction and impulse types, and combinations of each type may be found in a single turbine. Blade types are discussed in Sec. 6.6 but Fig. 6.2 describes how the velocity and pressure vary through impulse and reaction stages. The overall trend is a decreasing pressure with velocity recovery in the stator row or nozzle ring. A row of stator blades followed by a set of rotor blades is considered to be a stage and

Figure 6.1 160 MW axial flow steam turbine (*courtesy of Escher Wyss Ltd*)

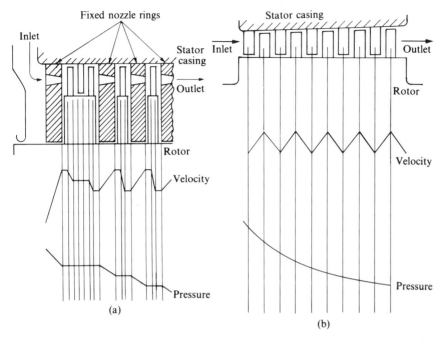

Figure 6.2 Pressure and velocity through impulse and reaction axial flow turbines: (a) impulse; (b) reaction

conditions through the stage will be examined in detail. The following assumptions will be made:

1. Flow conditions will be evaluated at the mean radius unless otherwise stated.
2. Blade height/mean radius is small, allowing two-dimensional flow theory to be used.
3. Radial velocities are zero.

6.2 TURBINE STAGE

A single turbine stage and velocity triangles are illustrated in Figs 6.3 and 6.4 respectively. The inlet to the stator blades is designated section 0, inlet to the rotor section 1 and outlet from the rotor section 2. This numbering system keeps the subscripts of the variables the same as in the case of the axial flow compressor. All flow angles are measured from the axial direction and care must be taken when reading steam turbine literature in which it is customary to measure flow angles from the direction of blade motion.

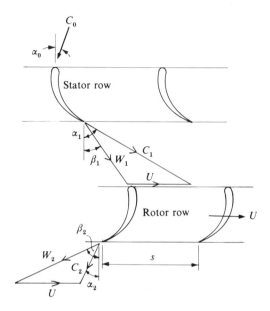

Figure 6.3 Velocity triangles for an axial flow gas (or steam) turbine stage

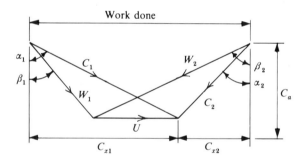

Figure 6.4 Superimposed velocity triangles

The gas leaves the stator blades with absolute velocity C_1 at angle α_1 and, by subtracting the blade velocity vector U, the relative velocity vector at entry to the rotor W_1 is determined. In moving across the rotor blade, the flow direction is changed, and the pressure reduced while the absolute velocity is decreased and the relative velocity increases. The gas leaves the blade tangentially at angle β_2 with relative velocity W_2. Vectorially subtracting the blade speed results in the absolute velocity C_2. This is now the inlet velocity to the next stator row at angle α_2, which for a normal stage equals C_0 at α_0 respectively. The two velocity triangles are conventionally superimposed upon each other as shown in Fig. 6.4, from which the energy transfer is given by

$$E = U(C_{x1} - C_{x2})/g$$

and since C_{x2} is in the negative x direction, the work done per unit mass flow is

$$Eg = W/m = U(C_{x1} + C_{x2}) \tag{6.1}$$

$$= U(W_{x1} + W_{x2}) \tag{6.2}$$

If $C_{a1} \neq C_{a2}$ there will be an axial thrust in the flow direction. However, we will assume that C_a is constant and therefore

$$W/m = UC_a(\tan \alpha_1 + \tan \alpha_2) \tag{6.3}$$

$$= UC_a(\tan \beta_1 + \tan \beta_2) \tag{6.4}$$

Equation (6.4) is often referred to as the diagram work per unit mass flow and associated with this is the diagram efficiency defined as

$$\text{Diagram efficiency} = \frac{\text{Diagram work done per unit mass flow}}{\text{Work available per unit mass flow}}$$

$$= U(W_{x1} + W_{x2})/\tfrac{1}{2}C_1^2 \tag{6.5}$$

The thermodynamic conditions of the gas through the stage are shown in Fig. 6.5, where the inlet conditions to the stator are at total pressure p_{00} and total enthalpy h_{00}. For adiabatic flow through the stator row or nozzle ring, $h_{00} = h_{01}$, but, owing to irreversibilities, the total pressure drops to p_{01} at stator outlet (rotor inlet). Expansion to p_{02} and total enthalpy h_{02} takes place in the rotor row and, as shown in Eq. (5.6), $h_{01\text{rel}} = h_{02\text{rel}}$. The work done per unit mass flow by the gas is given by

$$W/m = h_{00} - h_{02} = h_{01} - h_{02} \tag{6.6}$$

or

$$W/m = C_p(T_{01} - T_{02}) \tag{6.7}$$

Substituting for W/m from Eq. (6.4),

$$C_p(T_{01} - T_{02}) = UC_a(\tan \beta_1 + \tan \beta_2) \tag{6.8}$$

It should be noted that the work done factor λ is not used in Eq. (6.8). This is because, in a gas or steam turbine, flow through the blade passages is accelerating as opposed to decelerating. Flow in the compressor and the effect of boundary-layer growth in the former are therefore negligible. For a normal stage in which $C_0 = C_2$, the static temperature drop across the stage equals the total temperature drop.

The turbine stage total-to-total isentropic efficiency $\eta_{t(t-t)}$ is defined as

$$\eta_{t(t-t)} = \frac{\text{Actual work done by the gas}}{\text{Isentropic work done}}$$

$$= (T_{00} - T_{02})/(T_{00} - T_{02\text{ss}}) \tag{6.9}$$

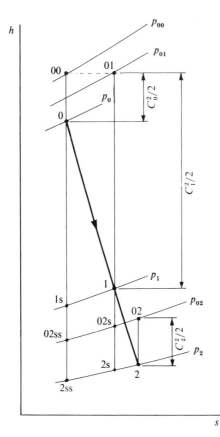

Figure 6.5 Mollier chart for expansion through an axial flow steam or gas turbine stage

Thus

$$T_{00} - T_{02} = \eta_{t(t-t)} T_{00}(1 - T_{02ss}/T_{00})$$

$$T_{00} - T_{02} = \eta_{t(t-t)} T_{00}[1 - (p_{02}/p_{00})^{(\gamma - 1)/\gamma}] \qquad (6.10)$$

6.3 STATOR (NOZZLE) AND ROTOR LOSSES

Before considering the losses occurring in the rotor and stator rows, it will be worth while distinguishing between two isentropic efficiencies commonly used in axial flow turbine work. The first has already been defined in Eq. (6.9) where the temperature limits are taken between total temperatures, and hence, defined in this way, the efficiency is the total-to-total efficiency. This form is used when the kinetic energy at the outlet of the stage is utilized for producing work. Such a case would be the propelling nozzle of a turbojet exhaust and any intermediate stage of a multi-stage turbine where the leaving kinetic energy is used in the following stage. The other efficiency is the total-to-static efficiency, which is used when the leaving kinetic energy is wasted, thus not contributing to the work output. It is defined as

$$\eta_{t(t-s)} = (h_{00} - h_{02})/(h_{00} - h_{2ss}) \qquad (6.11)$$

Considering the total-to-total efficiency of a normal stage,

$$\eta_{t(t-t)} = (h_{00} - h_{02})/(h_{00} - h_{02ss})$$
$$= (h_0 - h_2)/(h_0 - h_{2ss})$$

For a normal stage $C_0 = C_2$ and $\alpha_0 = \alpha_2$, and upon rearranging

$$\eta_{t(t-t)} = (h_0 - h_2)/[(h_0 - h_2) + (h_2 - h_{2s}) + (h_{2s} - h_{2ss})] \qquad (6.12)$$

But consider now the slope of a constant-pressure line on the Mollier chart (Fig. 6.5). Since $dh = T\,ds$ then the slope along a constant-pressure line is given by

$$(\partial h/\partial s)_p = T \qquad (6.13)$$

and therefore for a finite change of enthalpy Δh at constant pressure

$$\Delta h \approx T\Delta s$$

Therefore

$$(h_{2s} - h_{2ss}) \approx T_2(s_{2s} - s_{2ss}) \qquad (6.14)$$

and

$$(h_1 - h_{1s}) \approx T_1(s_1 - s_{1s})$$

But examination of Fig. 6.5 shows

$$(s_{2s} - s_{2ss}) = (s_1 - s_{1s})$$

Therefore substituting into Eq. (6.14) gives

$$(h_{2s} - h_{2ss}) = (T_2/T_1)(h_1 - h_{1s}) \qquad (6.15)$$

For both the stator (nozzle) and blade rows, dimensionless loss coefficients may be defined in two ways. For the nozzle

$$\zeta_N = (h_1 - h_{1s})/0.5C_1^2 \quad \text{or} \quad Y_N = (p_{00} - p_{01})/(p_{01} - p_1) \qquad (6.16)$$

and for the rotor

$$\zeta_R = (h_2 - h_{2s})/0.5W_2^2 \quad \text{or} \quad Y_R = (p_{01rel} - p_{02rel})/(p_{02rel} - p_2) \qquad (6.17)$$

The value of Y or ζ in the stator and rotor represents the percentage drop of energy due to friction in the blades, which results in a total pressure and static enthalpy drop across the blades. Typical losses are of the order of 10–15 per cent but can be lower for very low values of flow coefficient.

Substituting for the blade loss coefficients into Eq. (6.12), we get

$$\eta_{t(t-t)} = (h_0 - h_2)/[(h_0 - h_2) + W_2^2\zeta_R/2 + (T_2/T_1)\zeta_N C_1^2/2]$$
$$= \{1 + [\zeta_R W_2^2 + \zeta_N C_1^2(T_2/T_1)]/[2(h_0 - h_2)]\}^{-1} \qquad (6.18)$$

If the exit velocity is not utilized, then following a similar procedure for the analysis, the total-to-static efficiency becomes

$$\eta_{t(t-s)} = \{1 + [\zeta_R W_2^2 + \zeta_N C_1^2 (T_2/T_1) + C_0^2]/[2(h_0 - h_2)]\}^{-1} \quad (6.19)$$

As a starting point for a calculation, since T_2 and T_1 are not very different, the ratio (T_2/T_1) is often taken as unity. Typical values of ζ are given by

$$\zeta = 0.04 + 0.06(\varepsilon/100)^2 \quad (6.20)$$

where the deflection angle ε is $\varepsilon = \alpha_0 + \alpha_1$ for the stator (nozzle) and $\varepsilon = \beta_1 + \beta_2$ for the rotor.

Equation (6.20) applies only for a Reynolds number of 10^5, based on the characteristic velocity C_1 at the stator outlet and the characteristic hydraulic diameter d_H defined as

$$d_H = \frac{4 \times \text{Flow area at stator outlet}}{\text{Wetted perimeter at stator outlet}}$$

With reference to Fig. 6.6,

$$d_H = (4sl \cos \alpha_1)/[2(s \cos \alpha_1 + 1)]$$
$$= (2 sl \cos \alpha_1)/(s \cos \alpha_1 + 1)$$

The ratio of maximum blade thickness to chord (t_{max}/c) is 0.2 and the blade aspect ratio (span/chord) based on the axial chord (i.e. $c \cos \alpha_\infty$) is 3. For conditions other than the above, for instance if the aspect ratio is different, the following empirical equations may be used:

For rotors

$$1 + \zeta_1 = (1 + \zeta_{nom})[0.975 + (0.075c \cos \alpha_\infty)/l] \quad (6.21)$$

For stators (nozzles)

$$1 + \zeta_1 = (1 + \zeta_{nom})[0.933 + (0.021c \cos \alpha_\infty)/l] \quad (6.22)$$

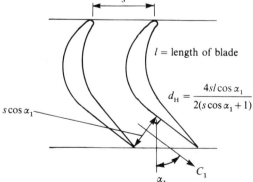

$$d_H = \frac{4sl \cos \alpha_1}{2(s \cos \alpha_1 + 1)}$$

l = length of blade

Figure 6.6 The hydraulic diameter

where ζ_1 and ζ_{nom} are the loss coefficients at the new aspect ratio and at an aspect ratio of 3 respectively.

Also, in Eqs (6.21) and (6.22) l is the blade length, while a Reynolds number of 10^5 is still assumed. Should the Reynolds number not be 10^5, a further correction may be made as follows. The Reynolds number for the flow is given by

$$Re = (2\rho_1 s l C_1 \cos \alpha_1)/[\mu(s \cos \alpha_1 + 1)] \tag{6.23}$$

and if ζ_1 is the loss coefficient at a Reynolds number of 10^5, then at the new Reynolds number, the corrected loss coefficient is given by

$$\zeta_2 = (10^5/Re)^{0.25} \zeta_1 \tag{6.24}$$

This empirical method of loss prediction is based on the Soderberg correlation[6] and is derived from the analysis of a large number of turbine characteristics. The error over a wide range of Reynolds numbers is as low as 3 per cent. Nozzle loss coefficients obtained from a large number of turbine tests are typically 0.09 and 0.05 for the rotor and stator rows respectively. The total-to-total efficiency is in the region of 90 per cent, the variation with blade loading and flow coefficient being plotted in Fig. 6.7, where it is seen that low ψ and ϕ give high stage efficiencies.

6.3.1 Blade Loading Coefficient

The work capacity of the stage is expressed in terms of a temperature drop coefficient or blade loading coefficient

$$\psi = W/mU^2 = C_p(T_{01} - T_{02})/U^2 \tag{6.25}$$

which from Eq. (6.4) may also be written as

$$\psi = C_a(\tan \beta_1 + \tan \beta_2)/U$$
$$= \phi(\tan \beta_1 + \tan \beta_2) \tag{6.26}$$

where ϕ is the flow coefficient.

The implication of a low flow coefficient is that frictional losses are reduced in the stage since C_a is low, but for a given mass flow rate the annulus flow area would be large. Low values of ψ imply a small amount of work done per stage, and therefore for a required power output, a large number of stages is required.

In stationary industrial power plants where the specific fuel consumption is of prime importance, a large-diameter, relatively long turbine, of low flow coefficient and low blade loading, giving a high efficiency, would probably be accepted. However, gas turbines used in aircraft propulsion have minimum weight and a small frontal area as prime considerations. This means using

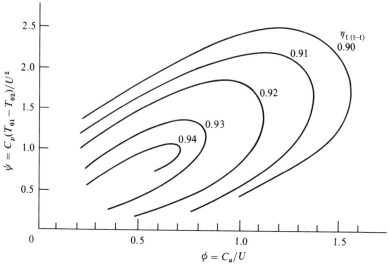

Figure 6.7 Blade loading coefficient versus flow coefficient

higher values of flow coefficient and blade loading factor to give a shorter compact turbine but also, as a consequence, accepting a lower efficiency (Fig. 6.7).

6.4 REACTION RATIO

The reaction ratio has more meaning in the case of an axial flow turbine than for an axial flow compressor where it is usually set at 50 per cent for the stage. The turbine reaction ratio can vary widely from 100 per cent to 0, the implications of which are described in this section.

The reaction ratio is

$$R = \frac{\text{Static enthalpy drop across rotor}}{\text{Static enthalpy drop across stage}}$$

$$= (h_1 - h_2)/(h_0 - h_2)$$
$$= (h_1 - h_2)/[(h_{00} - C_0^2/2) - (h_{02} - C_2^2/2)]$$

But for a normal stage $C_0 = C_2$ and since $h_{00} = h_{01}$ in the nozzle, then

$$R = (h_1 - h_2)/(h_{01} - h_{02}) \qquad (6.27)$$

Remembering that $h_{01\text{rel}} = h_{02\text{rel}}$, then

$$h_{01\text{rel}} - h_{02\text{rel}} = (h_1 - h_2) + (W_1^2 - W_2^2)/2 = 0$$

and substituting for $(h_1 - h_2)$ in Eq. (6.27),

$$R = (W_2^2 - W_1^2)/[2(h_{01} - h_{02})]$$
$$= (W_2^2 - W_1^2)/[2U(C_{x1} + C_{x2})] \tag{6.28}$$

If it is assumed that the axial velocity C_a is constant through the stage then

$$R = (W_{x2}^2 - W_{x1}^2)/[2U(U + W_{x1} + W_{x2} - U)]$$
$$= (W_{x2} - W_{x1})(W_{x2} + W_{x1})/[2U(W_{x1} + W_{x2})]$$
$$= C_a(\tan \beta_2 - \tan \beta_1)/2U$$
$$= \phi(\tan \beta_2 - \tan \beta_1)/2 \tag{6.29}$$

Equation (6.29) can be rearranged into a second form,

$$R = (C_a \tan \beta_2)/2U - [(C_a \tan \alpha_1)/2U - U/2U]$$
$$= 0.5 + C_a(\tan \beta_2 - \tan \alpha_1)/2U \tag{6.30}$$

and a third form is given by substituting for

$$\tan \beta_2 = (U + C_{x2})/C_a = (\tan \alpha_2 + U/C_a)$$

into Eq. (6.30), giving

$$R = 0.5 + C_a(\tan \alpha_2 + U/C_a - \tan \alpha_1)/2U$$
$$= 1 + C_a(\tan \alpha_2 - \tan \alpha_1)/2U \tag{6.31}$$

Inspection of Eq. (6.29) reveals that for zero reaction, $\beta_2 = \beta_1$ and $W_1 = W_2$. The gas conditions through the stage, and the accompanying velocity triangles, are shown in Fig. 6.8, where the velocity triangles are skewed to the left. Ideally, for reversible adiabatic flow, the points 1, 2 and 2s on the Mollier chart should coincide, and in that case no pressure drop occurs in the rotor. Now consider a pure impulse stage where, by definition, there is no pressure drop in the rotor. Figure 6.9 shows the Mollier chart for the pure impulse stage where, for reversible adiabatic flow, the points 1, 2 and 2s will coincide, and therefore with isentropic flow conditions prevailing

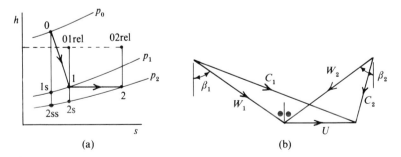

Figure 6.8 Zero reaction axial gas turbine

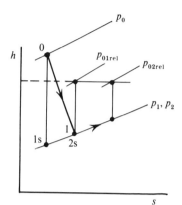

Figure 6.9 Pure impulse stage

the zero reaction stage is exactly the same as an impulse stage. However, when the flow is irreversible, the zero reaction and impulse stages are not the same and in fact an increase in enthalpy occurs in the rotor of the impulse stage, the implication being that the reaction is negative. For a negative reaction stage, the tendency will be for $W_2 < W_1$, thereby causing diffusion of the relative velocity vector in the rotor and a subsequent rise in pressure. This should be avoided since adverse pressure gradients causing flow separation on the blade surfaces can then occur. Figure 6.10 illustrates the Mollier chart for negative reaction.

With 50 per cent reaction, Eq. (6.30) requires that $\beta_2 = \alpha_1$, $\beta_1 = \alpha_2$. Also $C_1 = W_2$ and $C_2 = W_1$, resulting in the symmetrical velocity diagram of Fig. 6.11 with equal enthalpy drops in the stator and rotor.

For 100 per cent reaction, Eq. (6.31) gives $\alpha_1 = \alpha_2$ and $C_1 = C_2$, with the velocity diagram skewed to the right as illustrated in Fig. 6.12. Increasing the reaction ratio to greater than 1 gives rise to diffusion in the stator passages or nozzles with $C_1 < C_0$. This situation should also be avoided because of the likelihood of flow separation on the stator blade surfaces (Fig. 6.13).

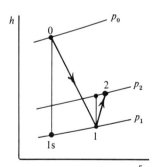

Figure 6.10 Negative reaction stage

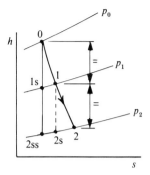

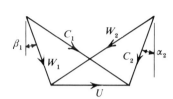

Figure 6.11 A 50 per cent reaction stage

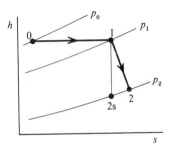

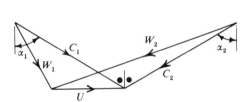

Figure 6.12 A 100 per cent reaction stage

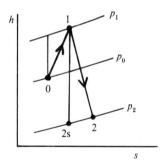

Figure 6.13 Stage expansion with reaction more than 100 per cent

6.5 EFFECT OF REACTION RATIO ON STAGE EFFICIENCY

Reaction and impulse blading find use in different types of machines. In steam turbines where, at the high-pressure end, pressure ratios across a stage would be very high, impulse blading is almost universally used since there is no pressure drop in the stage and therefore no tip leakage of the steam from one blade row to the next. At the low-pressure end of steam turbines and in gas turbines where the pressure ratios are more modest, reaction blading is employed. A choice of 50 per cent reaction means a sharing of the total expansion between the stator and rotor rows.

It was seen in Fig. 6.7 that, for a high total-to-total efficiency, the blade loading factor should be as low as possible, which implies as high a blade speed as possible, consistent with allowable blade stresses. However, the variation in total-to-total efficiency with slight variation in the blade loading coefficient is very small, this being true for a wide choice of reaction ratio. In contrast, the total-to-static efficiency is heavily dependent upon the reaction ratio and it can be shown that the total-to-static efficiency at a given blade loading may be optimized by choosing a suitable value of reaction.

6.6 BLADE TYPES

It has been noted in the previous section that steam turbines are usually impulse or a mixture of impulse and reaction stages whereas gas turbines tend to be always of the reaction type. The pressure ratio of steam turbines can be of the order of 1000:1 but for a gas turbine it is usually no more than 10:1. It is obvious that a very long steam turbine with many reaction stages would be required to reduce the pressure by a ratio of 1000:1, and even if the pressure drop per stage were made large to reduce the number of stages, blade tip leakage loss would still lead to very inefficient operation. Therefore reaction stages are used where the pressure drop per stage is low and also where the overall pressure ratio of the turbine is relatively low, as would be the case in an aero-engine, which may typically have only three or four reaction stages of or near 50 per cent reaction at the mean radius. The requirement of different types of stages has led to differing designs of blades for each type, and this section describes some of the pertinent points relevant to each design.

6.6.1 Reaction Blading

As described in Fig. 6.2, the pressure reduces through succeeding stator and rotor rows, the velocity being recovered as the pressure drops, and this necessitates a blade passage that is convergent towards the outlet, as in Fig. 6.14.

For 50 per cent reaction the stator and rotor blades will be the same, whereas zero reaction implies impulse rotor blades with constant cross-sectional area passages and no change in flow velocity. Reaction of 100 per cent implies that the stator blades are of the constant-area impulse type. It will be noted that the inlet angle β_1 for the reaction blade is almost zero while the profile of the back of the blade is almost linear. The form of

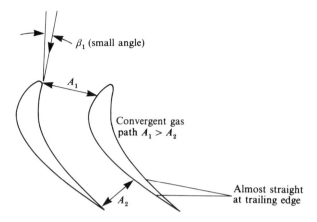

Figure 6.14 Reaction blading

Figure 6.15 Shrouded gas turbine reaction blades

the blade evolved empirically in steam turbine practice, and comparisons between blades designed in the late nineteenth century for the first steam turbine and those designed by the later aerodynamic theory showed only about a 4 per cent increase in maximum efficiency. However, efficiency increases of only 0.5 per cent or less are now very important when fuel costs are taken into account. Reaction blading is often shrouded at the tips, especially if the blades are long. This aids in preventing excessive vibration by tying the blades together and thus changing them from cantilevers to blades fixed at both ends (Fig. 6.15).

6.6.2 Impulse Blading

This type of blading has developed historically from steam turbine practice. It was noted in Sec. 6.1 that velocity triangles were drawn at the mean radius of the blades. Because of the tip speed variation along the length of the blade, incidence angles should change from root to tip, resulting in a twisted blade. However, it has been customary to ignore the twisted blade requirement and to have an untapered, untwisted blade of either 50 per cent reaction or zero reaction (impulse). The loss of efficiency of the straight blades is very small for the short blades that are used at the high-pressure end of the turbine. But for the much longer blades at the low-pressure end, larger losses can occur, and blades in this region are now designed on the free vortex principle instead of constant reaction ratio. For reversible flow through the rotor, impulse and zero reaction blading are the same.

Impulse blading is employed successfully at the high-pressure end of steam turbines. The velocity of the steam is increased in the convergent nozzle row to perhaps 800 m/s before entering the rotor blades and passing through them at constant pressure as in Fig. 6.2b. From Eq. (6.5), we can rearrange $(W_{x1} + W_{x2})$ as

$$W_{x1} + W_{x2} = W_1 \sin \beta_1 + W_2 \sin \beta_2 \qquad (6.32)$$

and putting $\beta_1 = \beta_2$ for impulse or zero reaction,

$$W_{x1} + W_{x2} = W_1 \sin \beta_1 (1 + W_R)$$
$$= (C_1 \sin \alpha_1 - U)(1 + W_R) \qquad (6.33)$$

where W_R is the relative velocity ratio W_2/W_1. Substituting Eq. (6.33) into Eq. (6.5)

$$\text{Diagram efficiency} = 2U(C_1 \sin \alpha_1 - U)(1 + W_R)/C_1^2$$
$$= 2[(U/C_1) \sin \alpha_1 - (U/C_1)^2](1 + W_R) \qquad (6.34)$$

For maximum diagram efficiency, differentiate Eq. (6.34) with respect to (U/C_1) and equate to zero. Then

$$0 = \sin \alpha_1 - 2U/C_1$$

or

$$U/C_1 = (\sin \alpha_1)/2 \qquad (6.35)$$

Equation (6.35) indicates that the nozzle angle α_1 should be as high as possible, the ideal being 90°. However, α_1 is limited by C_a, since the larger α_1, the smaller C_a becomes and therefore the longer is the blade length to accommodate the required mass flow rate. Typical nozzle angles are between 65° and 78°. The rotor blade passages are usually of constant-area symmetrical cross section, with inlet and outlet angles of 45° (β_1 and β_2) being typical (Fig. 6.16). The centres of curvature of the convex and concave surfaces of adjacent blades are then located at the same point to form parallel passages. Another design of impulse blading that has been developed is the convergent–divergent type. This design has been found useful in that

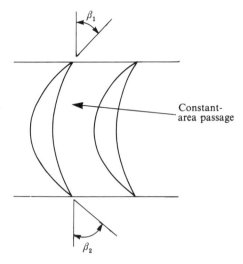

Figure 6.16 Impulse turbine blades

the chances of flow separation on the back convex surface of the blade are reduced by decreasing the radius of curvature of the convex surface. The slightly convergent passage aids in preserving uniform flow as the fluid flows into the bend, and although the diverging section causes diffusion of the flow, the net effect is beneficial when the turning angles are large and radii of blade curvature are small.

6.6.3 Free Vortex and Constant-Nozzle-Angle Design

It was pointed out in Sec. 6.6.2 that free vortex and other design criteria may be employed for the design of long blades, where constant-angle design would lead to low efficiencies.

As shown in Eq. (5.40), for constant stagnation enthalpy across the annulus ($\mathrm{d}h_0/\mathrm{d}r = 0$) and constant axial velocity ($\mathrm{d}C_a/\mathrm{d}r = 0$), then the whirl component of velocity C_x is inversely proportional to the radius and radial equilibrium is achieved. Now if we know conditions at the mean diameter and if subscript m denotes 'at the mean diameter', then at any nozzle blade radius r_1,

$$C_{x1}r_1 = r_1 C_{a1}\tan \alpha_1 = \text{const} \qquad \text{(free vortex condition)}$$

But C_{a1} is constant, and therefore at the nozzle outlet

$$\tan \alpha_1 = (r_{1m}/r_1)\tan \alpha_{1m}$$

and at the stage outlet when there is a whirl component

$$\tan \alpha_2 = (r_{2m}/r_2) \tan \alpha_{2m}$$

Another design criterion is that of constant-nozzle-angle design where the nozzle angle α_1 is constant as well as $dh_0/dr = 0$ and $dC_a/dr = 0$. This leads to the axial velocity distribution given by

$$C_{x1} r^{\sin^2 \alpha_1} = \text{const}$$

and since α_1 is constant, then C_{a1} is proportional to C_{x1} as well as $dh_0/dr = 0$. Therefore

$$C_{a1} r^{\sin^2 \alpha_1} = \text{const}$$

In many cases the change in vortex design has only a marginal effect on the performance of the blade while secondary losses may actually increase. It is left to the experience of the designer and the results of cascade test data to settle on a particular type of vortex flow.

6.7 AERODYNAMIC BLADE DESIGN

Cascade data similar to that discussed in Sec. 5.1 are used for modern turbine rotor blade design where the lift and drag coefficients are obtained from the cascade data curves. Pressure losses can then be determined and an estimation of the efficiency made. The drag coefficient must again be modified due to the blades actually being in annular form. Real boundaries exist at the hub and tip while the ideal flow pattern is disturbed by preceding and following blades. The drag coefficient is modified by tip clearance loss C_{DC} and secondary flow loss C_{DS} given by[7]

$$C_{DC} = nC_L^2(kc/ls) \tag{6.36}$$

and

$$C_{DS} = C_L^2 \lambda c/s \tag{6.37}$$

Here k is the clearance between the casing and blade tip while n is taken as 0.5 for unshrouded blades and 0.25 for tip-shrouded blades. The parameter λ is estimated from a functional relationship of the form

$$\lambda = f\left[\left(\frac{\text{Blade outlet area normal to flow}}{\text{Blade inlet area normal to flow}}\right)^2 \bigg/ \left(1 + \frac{\text{Hub radius}}{\text{Tip radius}}\right)\right]$$

Hence for the turbine

$$C_{DT} = C_D \text{ (from cascade data)} + C_{DC} + C_{DS} \tag{6.38}$$

Further information on blade design may be obtained from specialized texts, detailed design procedures being outside the scope of this book.

6.8 MULTI-STAGE GAS TURBINES

When the multi-stage compressor was discussed in Sec. 5.8, the polytropic or small-stage efficiency was made equal to the stage isentropic efficiency, and an initial estimate of the expected overall pressure ratio was obtained. In multi-stage gas turbines a similar procedure could be followed, but since there are probably only three or four stages at the most, little extra work is necessary to carry out a calculation stage by stage to determine the temperatures and pressures through the turbine, the outlet conditions of the preceding stage becoming the inlet conditions for the following stage. Once the overall temperature and pressure drops have been established, the isentropic efficiency can be calculated.

The performance characteristics are usually drawn in terms of the mass flow parameter $mT_{01}^{1/2}/p_{01}$ and efficiency η_t versus overall pressure ratio p_{01}/p_{0II} at differing speeds given by the parameter $N/T_{01}^{1/2}$. All speed curves are seen to be grouped closely together, merging into a single line at a maximum

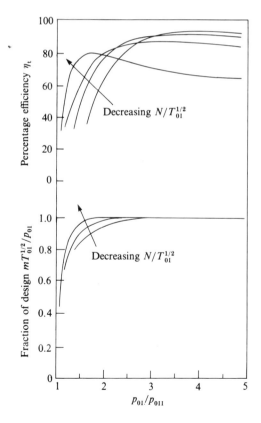

Figure 6.17 Characteristic curves for an axial flow gas turbine

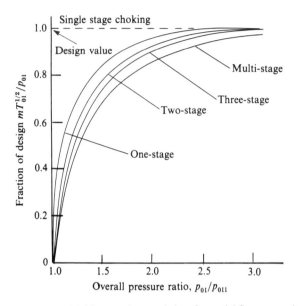

Figure 6.18 Multi-stage characteristics of an axial flow gas turbine

mass flow value (Fig. 6.17). This is the choked condition and is caused by velocities in the nozzle or at exit from the rotor becoming sonic.

Since the mass flow characteristics are grouped so closely together, even towards the lower pressure ratios, a mean curve through all the speed curves is often drawn and taken to be representative of that turbine. The isentropic efficiency remains within a narrow band for a wide range of pressure ratios (once choking has occurred) and speed, implying that the gas turbine is a very flexible machine as far as matching to a compresser is concerned, the compressor as we have seen in Sec. 5.9 being very sensitive to off-design conditions. Therefore, if the compressor design speed is different from the turbine design speed, the turbine efficiency will be little changed from its maximum value when running at the off-design speed. This flexibility of operation is due to a wide range of incidence angle being possible without incurring high rotor blade pressure loss coefficients and can be observed from cascade data.

Finally for this section, Fig. 6.18 shows the effect of increasing the number of stages. The limiting mass flow for the single-stage turbine occurs at a pressure ratio just in excess of 2, and is due to choking in the stator passages. The choking pressure ratio increases, however, with an increase in the number of stages, but for a given pressure ratio, the swallowing capacity (mass flow) decreases as the number of stages increases. The curve for the multi-stage turbine closely follows the ellipse law

$$mT_{01}^{1/2}/p_{01} = k[1 - (p_{011}/p_{01})^2]^{1/2} \qquad (6.39)$$

where k is a constant

EXERCISES

The following constants should be used, except where otherwise stated:
   Specific heat of turbine gas at constant pressure, $C_p = 1148 \, \text{J/kg K}$
   Ratio of specific heats, $\gamma = 1.333$, gas constant $R = 287 \, \text{J/kg K}$

6.1 An impulse stage of a steam turbine is designed for a nozzle isentropic enthalpy drop of 58 kJ/kg with a mass flow rate of 0.3 kg/s. The steam leaving the nozzle enters the rotor blade passage at an angle of 75° measured from the axial flow direction and has its relative velocity reduced by 5 per cent in the blade passage. If the blade speed is 173 m/s and the velocity coefficient of the nozzle is 0.98, determine:
   (a) the blade inlet angle,
   (b) the power developed,
   (c) the tangential thrust on the blades and
   (d) the stage diagram efficiency.
Assume that the blading is symmetrical.

6.2 A single-wheel impulse steam turbine has equiangular rotor blades that develop 3.75 kW and produce a torque in the disc of 1.62 Nm at a mean radius of 132.5 mm. The rotor receives 0.014 kg/s of steam from nozzles inclined at 70° to the axial direction and the steam discharges from the wheel chamber in an axial direction. Find:
   (a) the blade angles,
   (b) the diagram efficiency,
   (c) the end thrust on the shaft and
   (d) the tangential thrust on the blades.

6.3 A 50 per cent reaction steam turbine running at 450 rpm develops 5 MW and has a steam mass flow rate of 6.5 kg/kWh. At a particular stage in the expansion the absolute pressure is 85 kPa at a steam dryness fraction of 0.94. If the exit angle of the blades is 70°, measured from the axial flow direction, and the outlet relative velocity of the steam is 1.3 times the mean blade speed, find the blade height if the ratio of rotor hub diameter to blade height is 14.

6.4 In a zero reaction gas turbine, the blade speed at the mean diameter is 290 m/s. Gas leaves the nozzle ring at an angle of 65° to the axial direction while the stage inlet stagnation temperature is 1100 K. The following pressures were measured at the various locations:

At nozzle entry, stagnation pressure	400 kPa
At nozzle exit, stagnation pressure	390 kPa
At nozzle exit, static pressure	200 kPa
At rotor exit, static pressure	188 kPa

Assuming that the magnitude and direction of velocities at entry and exit of the stage are the same, determine the stage total-to-total efficiency.

6.5 An axial flow turbine stage has a flow coefficient of 0.65, a constant axial velocity and the gas

leaves the stator blades at an angle of 65° to the axial direction. Calculate:
(a) the blade loading coefficient,
(b) the rotor blade relative flow angles,
(c) the degree of reaction and
(d) the total-to-total and total-to-static efficiencies.

6.6 A small axial flow gas turbine with equal stage inlet and outlet velocities has the following design data based on the mean diameter:

Inlet stagnation temperature, T_{00}	1100 K
Inlet stagnation pressure, p_{00}	350 kPa
Axial flow velocity (constant through stage), C_a	260 m/s
Mass flow, m	18 kg/s
Blade speed, U	350 m/s
Nozzle efflux angle, α_1	60°
Gas stage exit angle, α_2	12°

Calculate:
(a) the rotor blade gas angles,
(b) the degree of reaction, blade loading coefficient and power output and
(c) the *total* nozzle throat area required if the throat is situated at the nozzle outlet and the nozzle loss coefficient is 0.05.

6.7 A single-stage axial flow gas turbine with convergent nozzles has the following data:

Inlet stagnation temperature, T_{00}	1150 K
Inlet stagnation pressure, p_{00}	420 kPa
Pressure ratio, p_{00}/p_{02}	1.9
Stage stagnation temperature drop, $(T_{00} - T_{02})$	150 K
Mass flow, m	25 kg/s
Mean blade speed, U	340 m/s
Rotational speed, N	15 000 rpm
Flow coefficient, ϕ	0.75
Angle of gas leaving stage, α_2	12°

If the axial velocity remains constant and the gas velocities at inlet and outlet are the same determine at the mean radius:
(a) the blade loading coefficient and degree of reaction,
(b) the gas angles,
(c) the required nozzle throat area,
(d) the annulus area at nozzle inlet and outlet and at rotor outlet if $\zeta_N = 0.05$ and
(e) the height and radius ratio of the annulus at the above locations.

6.8 Using the data of exercise 6.7, if the turbine is of free vortex design, find the gas angles $\alpha_1, \beta_1, \alpha_2$ and β_2 at the root and tip of the rotor blades and the relative Mach number at the inlet to the blade tip and root.

6.9 Once again using the data of exercises 6.7 and 6.8, if the design is based on a constant nozzle angle, calculate the angle β_1 at the blade root and tip, and compare them with those of exercise 6.8.

6.10 The data below apply to a single-stage axial flow gas turbine designed on free vortex theory. The outlet velocity is axial and the turbine is designed with a constant annulus area through the stage.

Mass flow, m	30 kg/s
Inlet stagnation temperature, T_{00}	1225 K
Inlet stagnation pressure, p_{00}	800 kPa
Temperature drop, $(T_{00} - T_{02})$	160 K
Isentropic efficiency, $\eta_{t(t-t)}$	0.9
Mean blade speed, U_m	330 m/s
Rotational speed, N	15 000 rpm
Outlet velocity, C_2	390 m/s

Calculate:
  (a) the blade height and radius ratio of the annulus,
  (b) the axial velocity at nozzle exit if flow in the nozzle is isentropic and
  (c) the inlet relative Mach number at the rotor blade root radius.

SOLUTIONS

Exercise 6.1 (a) A nozzle velocity coefficient was defined in Eq. (3.7) and in terms of enthalpy may be redefined as

$$C_v = \frac{\text{Actual nozzle velocity}}{\text{Ideal nozzle velocity}}$$

Now across the nozzle, $h_{00} = h_{01}$, and since no work is done in the nozzle,

$$h_0 + \frac{C_0^2}{2} = h_1 + \frac{C_1^2}{2}$$

But in the steam chest at entry to the nozzle, C_0 is negligibly small compared to C_1, and therefore

$$C_1^2 = 2(h_1 - h_2)$$

The ideal nozzle velocity will be when the flow through the nozzle is isentropic, that is when $(h_0 - h_1) \equiv (h_0 - h_{1s})$. Thus ideal $C_1^2 = 2(h_0 - h_{1s})$. Actual nozzle velocity

$$C_1 = (58 \times 10^3 \times 2)^{1/2} 0.98$$
$$= 333.8 \text{ m/s}$$

From the velocity diagrams of Figs 6.3 and 6.4

$$C_a = C_1 \cos \alpha_1$$
$$= 333.8 \cos 75°$$
$$= 86.4 \text{ m/s}$$

Also

$$U + W_{x1} = C_1 \sin \alpha_1$$
$$= 333.8 \sin 75°$$
$$= 322.4 \text{ m/s}$$

Whence

$$W_{x1} = 322.4 - 173$$
$$= 149.4 \, \text{m/s}$$

Therefore

$$\tan \beta_1 = W_{x1}/C_a$$
$$= \frac{149.4}{86.4}$$
$$\underline{\beta_1 = 60°}$$

(b) Relative inlet velocity

$$W_1 = C_a/\cos \beta_1$$
$$= \frac{86.4}{\cos 60°}$$
$$= 172.8 \, \text{m/s}$$

Therefore

$$W_2 = 0.95 \times 172.8$$
$$= 164.1 \, \text{m/s}$$
$$W_{x2} = W_2 \sin \beta_1 \qquad \text{since } \beta_1 = \beta_2$$
$$= 164.1 \sin 60°$$
$$= 142.1 \, \text{m/s}$$

From Eq. (6.2)

$$W/m = U(W_{x2} + W_{x1})$$

Power developed

$$W = 0.3 \times 173(142.1 + 172.8)$$
$$= \underline{16.3 \, \text{kW}}$$

(c) Tangential thrust

$$= m(W_{x2} + W_{x1})$$
$$= 0.3(142.1 + 172.8)$$
$$= \underline{94.47 \, \text{N}}$$

From Eq. (6.5)

(d) Diagram efficiency $= \dfrac{\text{Work done per unit mass flow}}{\text{Work available per unit mass flow}}$

$$= \frac{W/m}{C_1^2/2}$$

$$= \frac{16.3 \times 10^3 \times 2}{333.8^2 \times 0.3}$$

$$= 0.975$$

Exercise 6.2 (a) In the velocity diagram of Fig. 6.4, $C_a = C_2$ since discharge from the wheel chamber is axial. Then $W_{x2} = U$ at the mean radius.

$$\text{Torque} = mr(W_{x2} + W_{x1}) \qquad \text{at the mean radius}$$

$$W_{x2} + W_{x1} = \frac{1.62}{0.1325 \times 0.014}$$

$$= 873.3 \text{ m/s}$$

From Eq. (6.2)

$$W = mU(W_{x2} + W_{x1})$$

whence

$$U = \frac{3.75 \times 10^3}{0.014 \times 873.3}$$

$$= 306.7 \text{ m/s}$$

Now

$$C_{a1} = \frac{W_{x1} + W_{x2}}{\tan 70°}$$

$$= \frac{873.3}{\tan 70°}$$

$$= 317.9 \text{ m/s}$$

$$W_{x1} = 873.29 - 306.7$$

$$= 566.6 \text{ m/s}$$

Therefore

$$\tan \beta_1 = W_{x1}/C_{a1}$$

$$= \frac{566.6}{317.9}$$

$$\beta_1 = 60.7°$$

From Fig. 6.8

$$\beta_2 = 60.7°$$

(b) Also from the velocity diagrams

$$C_1^2 = (W_{x1} + W_{x2})^2 + C_{a1}^2$$

$$= 873.3^2 + 317.9^2$$

$$= 8.637 \times 10^5 \text{ m}^2/\text{s}^2$$

Equation (6.5) gives the diagram efficiency as

$$\text{Diagram efficiency} = \frac{U(W_{x2} + W_{x1})}{C_1^2/2}$$

$$= \frac{306.7 \times 873.3 \times 2}{863\,700}$$

$$\underline{= 0.62}$$

(c) If there is an axial thrust $C_{a1} \neq C_{a2}$

$$C_{a2} = \frac{W_{x2}}{\tan \beta_2}$$

$$= \frac{U}{\tan \beta_2} \qquad (U = W_{x2} \text{ since } C_2 \text{ is axial})$$

$$= \frac{306.7}{\tan 60.7°}$$

$$= 172.1 \text{ m/s}$$

$$\text{Axial thrust} = m(C_{a1} - C_{a2}) \qquad \text{(Eq. (1.23) in axial direction)}$$

$$= 0.014(317.9 - 172.1)$$

$$\underline{= 2.04 \text{ N}}$$

(d) Tangential thrust on blades $= m(W_{x2} + W_{x1}) \qquad$ (Eq. (1.23))

$$= 0.014 \times 873.3$$

$$\underline{= 12.23 \text{ N}}$$

Exercise 6.3 From the velocity diagram (Fig. 6.4)

$$W_2 = 1.3U$$

Now

$$C_a = W_2 \cos \beta_2$$

$$= 1.3U \cos 70°$$

$$= 0.445U \text{ m/s}$$

Now

$$U = \frac{\pi N D}{60} \qquad \text{at the mean diameter}$$

$$= \frac{2\pi N(D_h + h)}{60 \times 2}$$

where D_h is the rotor diameter at the hub and h is the blade height.

Substituting

$$C_a = \frac{0.445 \times 2\pi \times 450(14h + h)}{2 \times 60}$$

$$= 157.3\, h\, \text{m/s}$$

But the area of the annulus of flow $= \pi h(D_h + h)$, i.e.

$$A = 15\pi h^2$$

Now use steam tables, at a pressure of 85 kPa. At saturation the specific volume of the steam is $1.972\,\text{m}^3/\text{kg}$. At a dryness fraction of 0.94, then

$$\text{Specific volume of steam} = 1.972 \times 0.94$$
$$= 1.85\,\text{m}^3/\text{kg}$$

Now mass flow rate

$$m = \frac{5 \times 10^3 \times 6.5}{3600}$$

$$= 9.03\,\text{kg/s}$$

$$m = \frac{C_a A}{v} \qquad \text{where } v \text{ is specific volume}$$

Therefore

$$9.03 = \frac{157.3\, h \times 15\pi h^2}{1.85}$$

whence

$$h^3 = 2.254 \times 10^{-3}$$
$$h = 0.131\,\text{m}$$
$$\underline{\text{Blade height} = 0.131\,\text{m}}$$

Exercise 6.4 The condition of the gas through the stage is shown in Fig. 6.8 for zero reaction where it is noted that there is no temperature (or enthalpy) drop through the rotor.

For zero reaction

$$h_1 = h_2 \qquad \text{from Eq. (6.27)}$$

Also in the rotor

$$h_{01\text{rel}} = h_{02\text{rel}} \qquad \text{from Eq. (5.6)}$$

Therefore

$$W_1 = W_2 \qquad \text{and} \qquad \beta_1 = \beta_2$$

The total-to-total efficiency is given by Eq. (6.12) for a normal stage, which may be rearranged in terms of temperature differences as

$$\eta_{t(t-t)} = \frac{T_0 - T_2}{(T_0 - T_2) + (T_2 - T_{2ss})}$$

and so the problem resolves itself into an evaluation of the above temperature differences.

Through the nozzle row $T_{00} = T_{01}$ and at a point

$$T_1 = T_{01}\left(\frac{p_1}{p_{01}}\right)^{(\gamma-1)/\gamma}$$

$$= 1100\left(\frac{200}{390}\right)^{(1.333-1)/1.333}$$

$$= 931 \text{ K}$$

Then

$$C_1^2 = 2C_p(T_{01} - T_1)$$
$$= 2 \times 1148(1100 - 931)$$
$$= 388\,024$$
$$C_1 = 622.9 \text{ m/s}$$

Now

$$C_{a1} = C_1 \cos \alpha_1$$
$$= 622.9 \cos 65°$$
$$= 263.25 \text{ m/s}$$

and

$$C_{x1} = C_1 \sin \alpha_1$$
$$= 622.9 \sin 65°$$
$$= 564.5 \text{ m/s}$$

Therefore

$$W_{x1} = C_{x1} - U$$
$$= 564.5 - 290$$
$$= 274.5 \text{ m/s}$$

Now

$$C_{x2} = W_{x2} - U$$

and since we have zero reaction then $W_{x1} = W_{x2}$. So

$$C_{x2} = 274.5 - 290$$
$$= -15.5 \text{ m/s}$$

The negative sign indicates that there is a whirl velocity in the opposite direction to blade motion.

Since the stage is normal, $C_2 = C_0$, and assuming constant C_a through the stage

$$C_2^2 = C_a^2 + C_{x2}^2$$
$$= 263.25^2 + 15.5^2$$
$$= 69\,540\,\text{m}^2/\text{s}^2$$

Thus

$$T_0 - T_2 = T_{00} - \frac{C_0^2}{2C_p} - T_2$$

$$= 1100 - \left(\frac{69\,540}{2 \times 1148}\right) - 931 \qquad \text{since } T_1 = T_2$$

$$= 138.7\,\text{K}$$

Using isentropic relationships

$$\frac{T_{2ss}}{T_{00}} = \left(\frac{p_2}{p_{00}}\right)^{(\gamma - 1)/\gamma}$$

$$T_{2ss} = 1100 \left(\frac{188}{400}\right)^{0.25}$$

$$= 910.8\,\text{K}$$

Therefore

$$T_2 - T_{2ss} = 931 - 910.8$$
$$= 20.2\,\text{K}$$

Substituting for these temperature differences gives

$$\eta_{t(t-t)} = \frac{138.7}{138.7 + 20.2}$$

$$= 0.873$$

Exercise 6.5 (a) From Eq. (6.26), the blade loading coefficient is

$$\psi = \phi(\tan \beta_1 + \tan \beta_2)$$
$$= \phi(\tan \alpha_1 + \tan \alpha_2)$$

but in this case $\alpha_2 = 0$ since at outlet the flow is axial. Therefore

$$\psi = 0.65 \tan 65°$$
$$= 1.394$$

(b) From the velocity triangles

$$\tan \beta_2 = U/C_a \qquad \text{(if } C_2 \text{ is axial)}$$

$$= \frac{1}{0.65}$$

Blade outlet angle $\beta_2 = 57°$

$$\tan \beta_1 = \frac{\psi}{\phi} - \tan \beta_2$$

$$= \frac{1.394}{0.65} - \tan 57°$$

Blade inlet angle $\beta_1 = 31.2°$

(c) From Eq. (6.30)

$$R = 0.5 + 0.5\phi(\tan \beta_2 - \tan \alpha_1)$$
$$= 0.5 + 0.5 \times 0.65(\tan 57° - \tan 65°)$$

Degree of reaction $= 0.303$

(d) Equation (6.18) gives the total-to-total efficiency in terms of the nozzle and rotor loss coefficients:

$$\eta_{t(t-t)} = \left(1 + \frac{\zeta_R W_2^2 + \zeta_N C_1^2 (T_2/T_1)}{2(h_0 - h_2)}\right)^{-1}$$

Using the Soderberg correlation of Sec. 6.3, for the stator (nozzle) row

$$\varepsilon_N = \alpha_0 + \alpha_1$$
$$= 0° + 65°$$
$$= 65°$$

For the rotor row

$$\varepsilon_R = \beta_1 + \beta_2$$
$$= 31.2° + 57°$$
$$= 88.2°$$

Therefore from Eq. (6.20), rotor loss coefficient

$$\zeta_R = 0.04 + 0.06\left(\frac{88.2}{100}\right)^2$$

$$= 0.0867$$

and nozzle loss coefficient

$$\zeta_N = 0.04 + 0.06\left(\frac{65}{100}\right)^2$$

$$= 0.0654$$

Also

$$W_2^2 = C_a^2 \sec^2 \beta_2$$

and
$$C_1^2 = C_a^2 \sec^2 \alpha_1$$

and from Eq. (6.6)

$$
\begin{aligned}
h_0 - h_2 &= U(C_{x1} + C_{x2}) && \text{(since } C_0 = C_2) \\
&= U C_{x1} && \text{(no whirl at exit)} \\
&= U C_a \tan \alpha_1
\end{aligned}
$$

Therefore upon substituting

$$\eta_{t(t-t)} = \left(1 + \frac{0.0867 \sec^2 57° + 0.0654 \sec^2 65°(T_2/T_1)}{2 \tan 65°/0.65}\right)^{-1}$$

Putting $T_2/T_1 = 1$,

$$
\begin{aligned}
\eta_{t(t-t)} &= 0.909 \\
&= 90.9\%
\end{aligned}
$$

Equation (6.19) gives the total-to-static efficiency with $C_0 = C_a$

$$
\begin{aligned}
\eta_{t(t-s)} &= \left(1 + \frac{\zeta_R W_2^2 + \zeta_N C_1^2(T_2/T_1) + C_0^2}{2(h_0 - h_2)}\right)^{-1} \\
&= \left(1 + \frac{\zeta_R \sec^2 \beta_2 + \zeta_N \sec^2 \alpha_1 + 1}{2 \tan \alpha_1/\phi}\right)^{-1} \\
&= \left(1 + \frac{0.0867 \sec^2 57° + 0.0654 \sec^2 65° + 1}{2 \tan 65°/0.65}\right)^{-1} \\
&= 0.799 \\
&= 79.9\%
\end{aligned}
$$

Exercise 6.6 (a) From the velocity triangles (Fig. 6.4)

$$
\begin{aligned}
C_{x1} &= C_a \tan \alpha_1 \\
&= 260 \tan 60° \\
&= 450.3 \text{ m/s} \\
C_{x2} &= C_a \tan \alpha_2 \\
&= 260 \tan 12° \\
&= 55.26 \text{ m/s}
\end{aligned}
$$

Hence

$$
\begin{aligned}
W_{x1} &= C_{x1} - U \\
&= 450.3 - 350 \\
&= 100.3 \text{ m/s}
\end{aligned}
$$

Therefore

$$\beta_1 = \tan^{-1}\left(\frac{W_{x1}}{C_a}\right)$$

$$= \tan^{-1}\left(\frac{100.3}{260}\right)$$

$$= 21.1°$$

Also

$$W_{x2} = U + C_{x2}$$

$$= 350 + 55.26$$

$$= 405.3 \text{ m/s}$$

Therefore

$$\beta_2 = \tan^{-1}\left(\frac{W_{x2}}{C_a}\right)$$

$$= \tan^{-1}\left(\frac{405.3}{260}\right)$$

$$= 57.3°$$

(b) From Eq. (6.29), reaction

$$R = \frac{\phi}{2}(\tan\beta_2 - \tan\beta_1)$$

$$= \frac{260}{2 \times 350}(\tan 57.3° - \tan 21.1°)$$

$$= 0.435$$

Blade loading coefficient is given by Eq. (6.26)

$$\psi = \frac{C_a}{U}(\tan\beta_1 + \tan\beta_2)$$

[*Note*: Another definition for blade leading coefficient that is used is $\psi = C_p(T_{00} - T_{02})/\frac{1}{2}U^2$, which has a numerical value twice that of Eq. (6.26).]
Thus

$$\psi = \frac{260}{350}(\tan 21.1° + \tan 57.3°)$$

$$= 1.44$$

Power output

$$W = mU(C_{x1} + C_{x2})$$
$$= 18 \times 350(450.3 + 55.26)$$
$$= 3185 \, \text{kW}$$

(c) From velocity triangles

$$C_1 = C_a \sec \alpha_1$$
$$= 260 \sec 60°$$
$$= 520 \, \text{m/s}$$

To find the area of the nozzle throat we must find the density ρ_1. In Eq. (6.16) the nozzle enthalpy loss coefficient is

$$\zeta_N = \frac{C_p(T_1 - T_{1s})}{\frac{1}{2} C_1^2}$$

or

$$T_1 - T_{1s} = \frac{0.05 \times 0.5 \times 520^2}{1148}$$
$$= 5.89 \, \text{K}$$

Now in the nozzle $T_{01} = T_{00}$. That is, if the flow in the nozzle is adiabatic, then the stagnation temperature is constant. Now

$$T_1 = T_{01} - \frac{C_1^2}{2C_p}$$
$$= 1100 - \frac{520^2}{2 \times 1148}$$
$$= 982 \, \text{K}$$

Therefore

$$T_{1s} = 982 - 5.89$$
$$= 976.1 \, \text{K}$$

Thus

$$\frac{p_{00}}{p_1} = \left(\frac{T_{00}}{T_{1s}} \right)^{\gamma/(\gamma-1)}$$
$$= \left(\frac{1100}{976.1} \right)^4$$
$$= 1.61$$

$$p_1 = \frac{350}{1.61}$$
$$= 217 \, \text{kPa}$$

Then

$$\rho_1 = \frac{p_1}{RT_1}$$

$$= \frac{217 \times 10^3}{287 \times 982}$$

$$= 0.77 \text{ kg/m}^3$$

Mass flow rate

$$m = \rho_1 A_1 C_1$$

$$A_1 = \frac{18}{0.77 \times 520}$$

Nozzle outlet area $= 0.045 \text{ m}^2$

Exercise 6.7 (a) Substituting the data into Eq. (6.25) for the blade loading coefficient

$$\psi = \frac{C_p(T_{01} - T_{02})}{U^2}$$

$$= \frac{1148 \times 150}{340^2}$$

$$= 1.49$$

From the turbine velocity diagrams

$$U/C_a = \tan \beta_2 - \tan \alpha_2$$

or

$$\tan \beta_2 = \frac{1}{\phi} + \tan \alpha_2$$

$$= \frac{1}{0.75} + \tan 12°$$

$$\beta_2 = 57.1°$$

From Eqs (6.26) and (6.29),

$$\psi = \phi(\tan \beta_1 + \tan \beta_2)$$

and

$$R = \frac{\phi}{2}(\tan \beta_2 - \tan \beta_1)$$

from which

$$\tan \beta_2 = \frac{1}{2\phi}(\psi + 2R)$$

Therefore

$$\tan 57.1° = \frac{1}{2 \times 0.75}(1.49 + 2R)$$

whence reaction

$$R = 0.414$$

(b) Solving the above simultaneous equations for $\tan \beta_1$ gives

$$\tan \beta_1 = \frac{1}{2\phi}(\psi - 2R)$$

$$= \frac{1}{2 \times 0.75}(1.49 - 2 \times 0.414)$$

$$\beta_1 = 23.8°$$

Then

$$\tan \alpha_1 = \tan \beta_1 + \frac{1}{\phi}$$

$$= \tan 23.8° + \frac{1}{0.75}$$

$$\alpha_1 = 60.6°$$

(c) The nozzle throat area requires the density at the nozzle throat and the absolute velocity at that location. Since we have a simple convergent nozzle, the nozzle throat is at plane 1 in our notation.
From the velocity diagram

$$C_{a1} = U\phi$$
$$= 340 \times 0.75$$
$$= 255 \, \text{m/s}$$

and

$$C_1 = \frac{C_{a1}}{\cos \alpha_1}$$

$$= \frac{255}{\cos 60.6°}$$

$$= 519.5 \, \text{m/s}$$

Now

$$T_{01} - T_1 = \frac{C_1^2}{2C_p}$$

$$= \frac{519.5^2}{2 \times 1148}$$

$$= 117.6 \, \text{K}$$

From Eq. (6.16)

$$T_1 - T_{1s} = \frac{\zeta_N C_1^2}{2C_p}$$

$$= \frac{0.05 \times 519.5^2}{2 \times 1148}$$

$$= 5.87 \, \text{K}$$

and

$$T_{1s} = T_1 - 5.87$$

But

$$T_{01} = T_{00}$$

therefore

$$T_1 = 1150 - 117.6$$
$$= 1032.4 \, \text{K}$$

and hence

$$T_{1s} = 1032.4 - 5.87$$
$$= 1026.5 \, \text{K}$$

From isentropic relations between two points

$$\frac{p_{00}}{p_1} = \left(\frac{T_{00}}{T_{1s}}\right)^{\gamma/(\gamma-1)}$$

$$p_1 = 420\left(\frac{1026.5}{1150}\right)^4$$

$$= 266.6 \, \text{kPa}$$

Thus density

$$\rho_1 = \frac{p_1}{RT_1}$$

$$= \frac{272.8 \times 10^3}{287 \times 1032.4}$$

$$= 0.9 \, \text{kg/m}^3$$

$$\text{Nozzle throat area} = \frac{m}{\rho_1 C_1}$$

$$= \frac{25}{0.9 \times 519.5}$$

$$= 0.0534 \, \text{m}^2$$

(d) Annulus area at plane 1 is

$$A_1 = \frac{m}{\rho_1 C_{a1}}$$

$$= \frac{25}{0.9 \times 255}$$

Annulus area $A_1 = 0.109 \, \text{m}^2$

A similar procedure is followed for locations 0 and 2. Now

$$C_0 = C_2 = \frac{255}{\cos 12°}$$

$$= 260.7 \, \text{m/s}$$

and

$$\frac{C_0^2}{2C_p} = \frac{260.7^2}{2 \times 1148}$$

$$= 29.6 \, \text{K}$$

$$T_0 = T_{00} - \frac{C_0^2}{2C_p}$$

$$= 1150 - 29.6$$

$$= 1120 \, \text{K}$$

Therefore

$$\frac{p_{00}}{p_0} = \left(\frac{T_{00}}{T_0}\right)^{\gamma/(\gamma-1)}$$

$$p_0 = 420\left(\frac{1120}{1150}\right)^4$$

$$= 377.9 \, \text{kPa}$$

Hence

$$\rho_0 = \frac{p_0}{RT_0}$$

$$= \frac{378.4 \times 10^3}{287 \times 1120}$$

$$= 1.17 \, \text{kg/m}^3$$

Annulus area

$$A_0 = \frac{m}{\rho_0 C_a}$$

$$= \frac{25}{1.177 \times 255}$$

$$= 0.0838 \, \text{m}^2$$

At the stage outlet

$$T_{02} = T_{00} - (T_{00} - T_{02})$$

$$= 1150 - 150$$

$$= 1000 \, \text{K}$$

and

$$T_2 = T_{02} - \frac{C_2^2}{2C_p}$$

$$= 1000 - 29.6$$

$$= 970.4 \, \text{K}$$

Then

$$\frac{p_2}{p_{02}} = \left(\frac{T_2}{T_{02}} \right)^{\gamma/(\gamma-1)}$$

But

$$p_{02} = p_{00} \left(\frac{p_{02}}{p_{00}} \right)$$

$$= \frac{420}{1.9}$$

$$= 221.1 \, \text{kPa}$$

Then

$$p_2 = 221.1 \left(\frac{970.4}{1000} \right)^4$$

$$= 196.1 \, \text{kPa}$$

Hence

$$\rho_2 = \frac{196.1 \times 10^3}{287 \times 970.4}$$

$$= 0.704 \, \text{kg/m}^3$$

Therefore annulus area

$$A_2 = \frac{m}{\rho_2 C_a}$$

$$= \frac{25}{0.704 \times 255}$$

$$= 0.139 \, \text{m}^2$$

(e) If h is denoted as the blade height and r is the hub radius while R is the tip radius, then if the mean radius condition is denoted by subscript m

$$U_m = \frac{\pi N D_m}{60}$$

$$D_m = \frac{340 \times 60}{\pi \times 15\,000}$$

$$= 0.433 \, \text{m}$$

Now annulus area

$$A = \pi D_m h = \frac{60 U_m h}{N}$$

Then

$$h = \frac{AN}{60 U_m}$$

$$= 0.735 \, A$$

and so

$$h_0 = 0.061 \, \text{m}; \, h_1 = 0.080 \, \text{m}; \, h_2 = 0.102 \, \text{m}$$

Also

$$\frac{R}{r} = \frac{r_m + (h/2)}{r_m - (h/2)}$$

and so

$$(R/r)_0 = 1.328; \, (R/r)_1 = 1.45; \, (R/r)_2 = 1.61$$

Exercise 6.8 At the root

$$r = r_m - (h/2)$$

At the tip

$$R = r_m + (h/2)$$

Therefore

$$\left(\frac{r}{r_m}\right)_1 = 1 - \frac{h_1}{2r_m}$$

$$= 0.81$$

$$\left(\frac{R}{r_m}\right)_1 = 1.185$$

$$\left(\frac{r}{r_m}\right)_2 = 0.764$$

$$\left(\frac{R}{r_m}\right)_2 = 1.236$$

For free vortex design Sec. 6.6.3 shows that at the root

$$\tan \alpha_{1r} = \left(\frac{r_m}{r}\right)_1 \tan \alpha_{1m}$$

$$= 1.23 \tan 60.6°$$

$$\alpha_{1r} = 65.5°$$

and

$$\tan \alpha_{2r} = \left(\frac{r_m}{r}\right)_2 \tan \alpha_{2m}$$

$$= 1.31 \tan 12°$$

$$\underline{\alpha_{2r} = 15.5°}$$

At the tip

$$\tan \alpha_{1R} = 0.847 \tan 60.6°$$

$$\underline{\alpha_{1R} = 56.4°}$$

and

$$\tan \alpha_{2R} = 0.809 \tan 12°$$

$$\underline{\alpha_{2R} = 9.8°}$$

To find the relative gas angles, from the velocity triangles, at any radius,

$$\tan \beta_2 = \tan \alpha_2 + \frac{U}{C_a}$$

and

$$\tan \beta_1 = \tan \alpha_1 - \frac{U}{C_a}$$

But

$$\frac{U_m}{r_m} = \frac{U}{r}$$

Therefore substituting for U and $\tan \alpha$,

$$\tan \beta_2 = \left(\frac{r_m}{r}\right)_2 \tan \alpha_{2m} + \left(\frac{r}{r_m}\right)_2 \frac{U_m}{C_a}$$

and

$$\tan \beta_1 = \left(\frac{r_m}{r}\right)_1 \tan \alpha_{1m} - \left(\frac{r}{r_m}\right)_1 \frac{U_m}{C_a}$$

At the tip

$$\tan \beta_{2R} = \left(\frac{R}{r_m}\right)_2 \frac{U_m}{C_a} + \left(\frac{r_m}{R}\right)_2 \tan \alpha_{2m}$$

$$= \frac{1.236}{0.75} + \frac{\tan 12°}{1.236}$$

$$\underline{\beta_{2R} = 61.2°}$$

$$\tan \beta_{1R} = \frac{\tan 60.6°}{1.181} - \frac{1.181}{0.75}$$

$$\underline{\beta_{1R} = -4.0°}$$

At the root

$$\tan \beta_{2r} = \frac{\tan 12°}{0.764} + \frac{0.764}{0.75}$$

$$\underline{\beta_{2r} = 52.3°}$$

and

$$\tan \beta_{1r} = \frac{\tan 60.6}{0.81} - \frac{0.81}{0.75}$$

$$\underline{= 48°}$$

Summarizing the results:

	α_1	β_1	α_2	β_2
At tip	56.4°	-4.0°	9.8°	61.2°
At root	65.5°	48.0°	15.5°	52.3°
At mean radius	60.6°	23.8°	12.0°	57.1°

Blade angles at other points can be calculated to give the blade geometry over the whole span.

From exercise 6.7

$$W_1 = C_a \sec \beta_1$$

At the root

$$W_{1r} = 255 \sec \beta_{1r}$$
$$= 255 \sec 48°$$
$$= 381 \text{ m/s}$$

Also

$$C_{1r} = C_a \sec \alpha_{1r}$$
$$= 255 \sec 65.5°$$
$$= 615 \text{ m/s}$$

Thus

$$T_{1r} = T_{01} - \frac{C_{1r}^2}{2C_p}$$

$$= 1150 - \frac{615^2}{2 \times 1148}$$

$$= 985 \text{ K}$$

Thus Mach number

$$(M_{1\,\mathrm{rel}})_r = \frac{W_{1r}}{(\gamma R T_{1r})^{1/2}}$$

$$= \frac{381}{(1.33 \times 287 \times 985)^{1/2}}$$

$$\underline{\text{Mach number at root} = 0.664}$$

At the tip

$$W_{1R} = 255.7 \text{ m/s}$$
$$C_{1R} = 460.8 \text{ m/s}$$
$$T_{1R} = 1057.4 \text{ K}$$

At tip, Mach number

$$(M_{1\,\mathrm{rel}})_R = 0.402$$

It will be noted that the higher Mach number occurs at the blade root.

Exercise 6.9 The radial equilibrium equation is given by Eq. (5.37)

$$\frac{dh_0}{dr} = \frac{C_x^2}{r} + C_x \frac{dC_x}{dr} + C_a \frac{dC_a}{dr}$$

In plane 1, $dh_0/dr = 0$ as before and also if α_1 is constant then

$$\tan \alpha_1 = \frac{C_{x1}}{C_{a1}} = \text{const}$$

Then

$$\frac{dC_{a1}}{dr} = \frac{dC_{x1}}{dr} \cot \alpha_1$$

Equation (5.37) therefore becomes

$$0 = \frac{C_{x1}^2}{r} + C_{x1} \frac{dC_{x1}}{dr} + C_{x1} \cot^2 \alpha_1 \frac{dC_{x1}}{dr}$$

$$= (1 + \cot^2 \alpha_1) \frac{dC_{x1}}{dr} + \frac{C_{x1}}{r}$$

But

$$1 + \cot^2 \alpha_1 = \sin^2 \alpha_1$$

$$\frac{dC_{x1}}{C_{x1}} = -\sin^2 \alpha_1 \frac{dr}{r}$$

Integrating

$$\log_e C_{x1} = -\sin^2 \alpha_1 \log_e r + \text{const}$$

or

$$C_{x1} r^{\sin^2 \alpha_1} = \text{const} \qquad \text{(see Sec. 6.6.3)}$$

Now if α_1 is constant and $\tan \alpha_1 = C_{x1}/C_a$ then C_{x1} is proportional to C_a; therefore

$$C_a r^{\sin^2 \alpha_1} = \text{const}$$

At the mean radius $C_{am} = 255\,\text{m/s}$ and $r_{1m} = 0.2165\,\text{m}$, $\alpha_1 = 60.6°$. Therefore

$$\text{const} = 255 \times 0.2165^{\sin^2 60.6°}$$

$$= 79.83\,\text{m/s}$$

At the *root*

$$(C_{a1})_r = \frac{79.83}{(r_{1m} - h_1/2)^{0.759}}$$

$$= \frac{79.83}{(0.2165 - 0.08/2)^{0.759}}$$

$$= 297.77\,\text{m/s}$$

Thus

$$\phi_r = C_{a1}/U_1$$

$$= \frac{C_{a1}}{U_m} \left(\frac{r_m}{r}\right)_1$$

$$= \frac{297.77}{340} \times \frac{1}{0.81}$$

$$= 1.08 \approx 1$$

$$\tan \beta_{1r} = \tan \alpha_{1r} - \frac{1}{\phi_r}$$

$$= \tan 60.6 - \frac{1}{1.08} \qquad (\alpha_{1r} = \alpha_{1m} = \text{const})$$

$$\beta_{1r} = 40°$$

At the root $\beta_{1r} = 40°$

At the *tip*

$$(C_{a1})_R = \frac{79.83}{(r_{1m} + h_1/2)^{0.759}}$$

$$= \frac{79.83}{(0.2165 + 0.08/2)^{0.759}}$$

$$= 224.2 \, \text{m/s}$$

So

$$\phi_R = \frac{224.2}{340} \left(\frac{r_m}{R} \right)_1$$

$$= \frac{224.2}{340} \left(\frac{1}{1.18} \right)$$

$$= 0.559$$

$$\tan \beta_{1R} = \tan \alpha_{1R} - \frac{1}{\phi_R}$$

$$= \tan 60.6 - \frac{1}{0.559} \qquad (\alpha_{1R} = \alpha_{1m} = \text{const})$$

$$\beta_{1R} = -0.81°$$

At the $\beta_1 = -0.81°$

Exercise 6.10 (a) From Eq. (6.9)

$$T_{00} - T_{02ss} = \frac{T_{00} - T_{02}}{\eta_{t(t-t)}}$$

$$= \frac{160}{0.9}$$

$$= 177.77 \, \text{K}$$

Therefore

$$T_{02ss} = 1225 - 177.77$$

$$= 1047.2 \, \text{K}$$

From isentropic relationships

$$p_{02} = p_{00}\left(\frac{T_{02ss}}{T_{00}}\right)^{\gamma/(\gamma-1)}$$

$$= 800\left(\frac{1047.2}{1225}\right)^4$$

$$= 427.3\,\text{kPa}$$

Now

$$T_2 = T_{02} - \frac{C_2^2}{2C_p}$$

$$= (1225 - 160) - \frac{390^2}{2 \times 1148}$$

$$= 998.7\,\text{K}$$

Then

$$p_2 = p_{02}\left(\frac{T_2}{T_{02}}\right)^{\gamma/(\gamma-1)}$$

$$= 427.3\left(\frac{998.7}{1065}\right)^4$$

$$= 330\,\text{kPa}$$

and

$$\rho_2 = \frac{p_2}{RT_2}$$

$$= \frac{330 \times 10^3}{287 \times 998.7}$$

$$= 1.153\,\text{kg/m}^3$$

From continuity

$$A_2 = \frac{m}{\rho_2 C_{a2}}$$

$$= \frac{30}{1.153 \times 390} \qquad \text{since } C_2 = C_a$$

$$= 0.0667\,\text{m}^2$$

At mean radius

$$r_\text{m} = \frac{U_\text{m}}{\omega}$$

$$= \frac{330 \times 60}{15\,000 \times 2\pi}$$

$$= 0.21\,\text{m}$$

Blade height

$$h = \frac{A_2}{2\pi r_m}$$

$$= \frac{0.0667}{2\pi \times 0.21}$$

$$= 0.0506 \text{ m}$$

Radius ratio

$$\frac{R}{r} = \frac{r_m + h/2}{r_m - h/2}$$

$$= \frac{0.21 + 0.0253}{0.21 - 0.0253}$$

$$= 1.27$$

(b) An iterative procedure must be followed to determine C_{a1}. This is set out below. The continuity equation must be satisfied; therefore at the nozzle outlet station 1,

$$m = \rho_1 C_{a1} A_1 \qquad A_1 = \text{const} \quad \text{and} \quad C_{a1} = \text{const}$$

As a first guess let

$$\rho_1 = \frac{p_{00}}{R T_{00}}$$

$$= \frac{800 \times 10^3}{287 \times 1225}$$

$$= 2.27 \text{ kg/m}^3$$

Then

$$C_{a1} = \frac{30}{2.27 \times 0.0667} \qquad (A_1 = A_2)$$

$$= 198 \text{ m/s}$$

Now remembering that $C_{x2} = 0$,

$$\frac{W}{m} = U(C_{x1} + C_{x2})$$

$$= U C_{a1} \tan \alpha_1$$

$$1148 \times 160 = 330 \times 198 \tan \alpha_1$$

from which

$$\tan \alpha_1 = \frac{1148 \times 160}{330 \times 198}$$

$$\alpha_1 = 70.4^\circ$$

Then

$$C_1 = \frac{C_{a1}}{\cos \alpha_1}$$

$$= \frac{198}{\cos 70.4°}$$

$$= 590.4 \, \text{m/s}$$

Now

$$T_1 = T_{00} - \frac{C_1^2}{2C_p}$$

$$= 1225 - \frac{590.4^2}{2 \times 1148}$$

$$= 1073 \, \text{K}$$

Using isentropic relations

$$p_1 = p_{00} \left(\frac{T_1}{T_{00}} \right)^{\gamma/(\gamma - 1)} \qquad \text{since } p_{00} = p_{01}$$

$$= 800 \left(\frac{1073}{1225} \right)^4$$

$$= 470.9 \, \text{kPa}$$

Thus

$$\rho_1 = \frac{p_1}{RT_1}$$

$$= \frac{470.9 \times 10^3}{287 \times 1073}$$

$$= 1.529 \, \text{kg/m}^3$$

This value is now used in a new iteration until C_{a1} and ρ_1 do not change. A table in set up below.

Iteration	1	2	3	4	5	6	7
ρ_1	2.27	1.53	1.44	1.42	1.42		
C_{a1}	200.2	294	312.3	316.7			
α_1	70.2	62.1	60.67	60.34			
C_1	590.4	628.3	637.6	640			
T_1	1073	1053	1047.8	1046.4			
p_1	470.9	436.6	428.2	426			
ρ_1	1.53	1.44	1.42	1.42			

At nozzle outlet axial velocity

$$\underline{C_{a1} = 316.7\,\text{m/s}}$$

(c) To find $(M_{1\text{rel}})_r$, where subscript r refers to the root radius,

$$\frac{U_r}{U_m} = \frac{r}{r_m}$$

$$U_r = 330 \times \frac{0.21 - 0.0253}{0.21}$$

$$= 290.2\,\text{m/s}$$

At the root

$$W_{x1} = \frac{C_p(T_{00} - T_{02})}{U} - U$$

$$= \frac{1148 \times 160}{290.2} - 290.2$$

$$= 342.7\,\text{m/s}$$

Then

$$W_1^2 = C_{a1}^2 + W_{x1}^2$$

$$= 316.7^2 + 342.7^2$$

$$W_1 = 466.6\,\text{m/s}$$

Acoustic velocity

$$a_1 = (\gamma R T_1)^{1/2}$$

$$= (1.333 \times 287 \times 1046.4)^{1/2}$$

$$= 632.7$$

$$M_{1\text{rel}} = W_1/a_1$$

$$= \frac{466.6}{632.7}$$

$$\underline{\text{Relative Mach number at root} = 0.74}$$

SEVEN

RADIAL FLOW GAS TURBINES

7.1 INTRODUCTION

The inward flow radial gas turbine is used for applications where the flow rate is very low, for example turbochargers for commercial (diesel) engines and fire pumps. They are very compact, the maximum diameter being about 0.2 m. Speeds are high, ranging from 40 000 to 180 0000 rpm. They are usually of the 90° type, the blades being perpendicular to the tangent at the rotor outer inlet periphery, and the gas after entering in the radial direction exits axially at outlet.

The turbine and its essential parts are shown in Fig. 7.1, where its similarity to the centrifugal compressor is noted, the difference being that the gas flow is in the opposite direction. Figure 7.1 shows that gas enters the scroll casing, the cross-sectional area of the scroll decreasing as the gas passes through it. This keeps the velocity at entry to the nozzle vanes constant as the gas is gradually drawn off on its circumferential path. The nozzle vanes are converging to increase the kinetic energy of the gas and they set the gas angle for entry into the rotor. This angle is usually about 70° (measured from the radial direction) but the vanes can be pivoted to allow for adjustment of the flow angle as the load changes. In some designs, there may be no vanes at all, but a passage similar to that of the vaneless diffuser of Fig. 2.17 is fitted (Fig. 7.2). A vaneless space exists between the outlet tip of the vanes and the rotor, this space being utilized by the gas for further flow adjustment and aiding in the reduction of vibratory disturbances within the turbine.

The rotor, which is usually manufactured of cast nickel alloy, has blades that are curved to change the flow from the radial to the axial direction. The

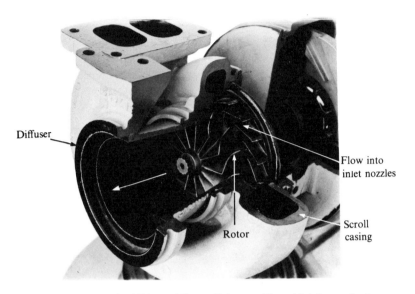

Figure 7.1 Elements of a 90° inward flow radial gas turbine with inlet nozzle ring

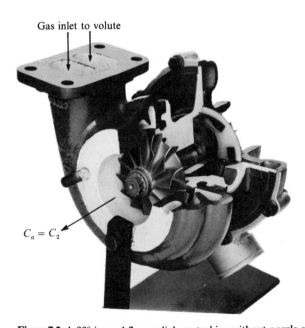

Figure 7.2 A 90° inward flow radial gas turbine without nozzle ring

shrouding for the blades is formed by the casing, and a diffuser can be fitted at the outlet, to reduce further the high kinetic energy at that point and thereby to increase the enthalpy drop across the rotor.

7.2 VELOCITY DIAGRAMS AND THERMODYNAMICS OF FLOW

The velocity triangles for this turbine are drawn in Fig. 7.3. The usual numbering system of 1 to 2 across the rotor will be used and therefore 0 indicates the point of entry to the nozzle vanes and 3 the diffuser outlet section. The thermodynamic path followed by the gas is shown on the Mollier chart of Fig. 7.4. In the nozzle no work is done, therefore $h_{00} = h_{01}$ although the total pressure drops from p_{00} to p_{01} because of irreversibilities. Thus

$$h_0 - h_1 = (C_1^2 - C_0^2)/2 \tag{7.1}$$

The work done per unit mass flow in the rotor is given by Euler's turbine equation (Eq. (1.24))

$$W/m = (U_1 C_{x1} - U_2 C_{x2}) \quad \text{(J/s)/(kg/s)} \tag{7.2}$$

If the whirl velocity is zero at exit then

$$W/m = U_1 C_{x1}$$

and for radial relative velocity at entry

$$W/m = U_1^2 \tag{7.3}$$

In more general terms, substituting for W/m

$$h_{01} - h_{02} = U_1 C_{x1} - U_2 C_{x2}$$

But it was shown in Sec 4.1.2 that the quantity I for a centrifugal compressor is

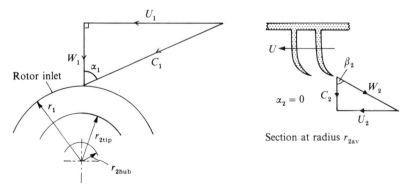

Figure 7.3 Velocity triangles for the 90° inward flow radial gas turbine

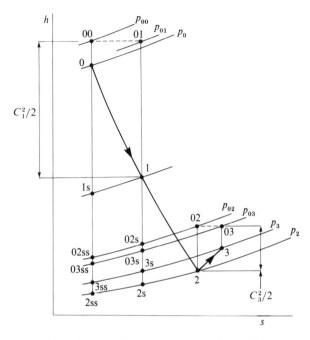

Figure 7.4 Mollier chart for expansion in a 90° inward flow radial gas turbine

given by

$$I = h_{0\text{rel}} - U^2/2 \qquad \text{where } I \text{ is a constant}$$

Therefore

$$h_1 - h_2 = [(U_1^2 - U_2^2) - (W_1^2 - W_2^2)]/2$$

If $C_{x2} = 0$, then $(W_2^2 - U_2^2) = C_2^2$ and

$$h_1 - h_2 = (U_1^2 - W_1^2 + C_2^2)/2 \tag{7.4}$$

In the diffuser $h_{02} = h_{03}$. Thus

$$h_3 - h_2 = (C_2^2 - C_3^2)/2 \tag{7.5}$$

If the losses in the diffuser are neglected, then $T_{03ss} = T_{02ss}$ and the total-to-total isentropic efficiency is given by

$$\eta_{t(t-t)} = (T_{00} - T_{02})/(T_{00} - T_{02ss}) \tag{7.6}$$

efficiencies being in the region of 80–90 per cent.

7.3 SPOUTING VELOCITY

If the gas passes through the turbine isentropically, then the isentropic total enthalpy drop is given by $(h_{00} - h_{02ss})$ if no diffuser is fitted and by $(h_{00} - h_{03ss})$

Table 7.1 Spouting velocities

	Total-to-total	Total-to-static
With diffuser	$C_s^2 = (h_{00} - h_{03ss})/2$	$C_s^2 = (h_{00} - h_{3ss})/2$
Without diffuser	$C_s^2 = (h_{00} - h_{02ss})/2$	$C_s^2 = (h_{00} - h_{2ss})/2$

if a diffuser is fitted. This energy change may be related to a kinetic energy, where the associated velocity term is known as the spouting velocity C_s. Thus four spouting velocities may be defined, as in Table 7.1, with and without a diffuser and for total-to-total or total-to-static conditions.

The appropriate definition would be used depending upon the efficiency being determined. It will be noted that for isentropic flow throughout

$$W/m = U_1^2 = C_s^2/2$$

or

$$U_1/C_s = 0.707 \tag{7.7}$$

In practice U_1/C_s lies in the range 0.68 to 0.7.

7.4 TURBINE EFFICIENCY

The method of determining the efficiency of an inward flow radial turbine is to determine the total-to-static efficiency in terms of loss coefficients for the nozzle and rotor, and then obtain a relationship for the total-to-total efficiency in terms of known turbine dimensions and the previously determined total-to-static efficiency. However, before doing that, a number of relationships that will be needed in the derivation of $\eta_{t(t-s)}$ are discussed. From Fig. 7.3

$$W_1 = U_1 \cot \alpha_1$$

$$C_2 = U_2 \cot \beta_2$$

and substituting for W_1 and C_2 into Eq. (7.4) yields

$$h_1 - h_2 = U_1^2[1 - \cot^2 \alpha_1 + (U_2/U_1)^2 \cot^2 \beta_2]/2$$

and putting $U_2/U_1 = r_2/r_1$

$$h_1 - h_2 = U_1^2[1 - \cot^2 \alpha_1 + (r_2/r_1)^2 \cot^2 \beta_2]/2 \tag{7.8}$$

Now $(h_1 - h_2) = C_p(T_1 - T_2) = \gamma R(T_1 - T_2)/(\gamma - 1)$ and Eq. (7.8) becomes

$$T_2/T_1 = 1 - U_1^2(\gamma - 1)[1 - \cot^2 \alpha_1 + (r_2/r_1)^2 \cot^2 \beta_2]/2\gamma R T_1$$

and putting $\gamma R T_1 = a_1^2$

$$T_2/T_1 = 1 - (U_1/a_1)^2(\gamma - 1)[1 - \cot^2 \alpha_1 + (r_2/r_1)^2 \cot^2 \beta_2]/2 \tag{7.9}$$

The total-to-static efficiency (without the diffuser) is given by

$$\eta_{t(t-s)} = (h_{00} - h_{02})/(h_{00} - h_{2ss})$$

$$= (h_{00} - h_{02})/[(h_{00} - h_{2s}) + (h_{2s} - h_{2ss})]$$

But

$$h_{00} = U_1^2 + h_{02} \qquad \text{(since } U_1^2 = W/m)$$

$$= U_1^2 + h_2 + C_2^2/2$$

Therefore

$$\eta_{t(t-s)} = U_1^2/[U_1^2 + C_2^2/2 + (h_2 - h_{2s}) + (h_{2s} - h_{2ss})] \qquad (7.10)$$

Defining a nozzle and rotor loss coefficient in terms of the enthalpy loss in each divided by the kinetic energy at outlet from each,

$$\zeta_N = (h_1 - h_{1s})/(0.5C_1^2) \qquad (7.11)$$

But from inspection of the Mollier chart

$$h_1 - h_{1s} \approx T_1(s_1 - s_{1s})$$

and

$$h_{2s} - h_{2ss} \approx T_2(s_{2s} - s_{2ss})$$

from which

$$h_1 - h_{1s} = (T_1/T_2)(h_{2s} - h_{2ss})$$

and substituting into Eq. (7.11) and rearranging gives

$$h_{2s} - h_{2ss} = (T_2/T_1)\zeta_N C_1^2/2 \qquad (7.12)$$

Similarly

$$\zeta_R = (h_2 - h_{2s})/(0.5W_2^2)$$

or

$$h_2 - h_{2s} = \zeta_R W_2^2/2 \qquad (7.13)$$

Now substituting Eqs (7.12) and (7.13) into Eq. (7.10)

$$\eta_{t(t-s)} = U_1^2/\{U_1^2 + [C_2^2 + W_2^2\zeta_R + C_1^2\zeta_N(T_2/T_1)]/2\}$$

$$= U_1^2/\{U_1^2 + [U_2^2 \cot^2\beta_2 + (U_2^2 \operatorname{cosec}^2\beta_2)\zeta_R$$
$$+ (U_1^2 \operatorname{cosec}^2\alpha_1)(T_2/T_1)\zeta_N]/2\}$$

$$= \{1 + [\zeta_N(T_2/T_1)\operatorname{cosec}^2\alpha_1 + (r_2/r_1)^2(\zeta_R \operatorname{cosec}^2\beta_2$$
$$+ \cot^2\beta_2)]/2\}^{-1} \qquad (7.14)$$

At the outlet section 2, two radii are possible, viz. at the shroud and at the hub. It is usual to take the average radius

$$r_{2av} = (r_{2hub} + r_{2tip})/2$$

while the temperature ratio (T_2/T_1) is obtained from Eq. (7.9).

However, substituting typical values of the variables into Eq. (7.9) yields (T_2/T_1) approximately equal to 1 and so it is often ignored and the approximate equation for total-to-static efficiency is given by

$$\eta_{t(t-s)} = \{1 + [\zeta_N \cosec^2 \alpha_1 + (r_{2av}/r_1)^2(\zeta_R \cosec^2 \beta_{2av} + \cot^2 \beta_{2av})]/2\}^{-1} \quad (7.15)$$

The total-to-total efficiency may be found from the following equation

$$(1/\eta_{t(t-t)}) = (1/\eta_{t(t-s)}) - [(r_{2av}/r_1)^2 \cot^2 \beta_{2av}]/2 \quad (7.16)$$

Loss coefficients usually lie in the following ranges for 90° inward flow turbines[8]

$$0.063 \leqslant \zeta_N \leqslant 0.235$$

$$0.384 \leqslant \zeta_R \leqslant 0.777$$

7.5 DIMENSIONLESS SPECIFIC SPEED

The inward flow radial gas turbine can be shown to have its maximum efficiency in a very narrow range of dimensionless specific speeds. However, whereas the volume flow rate through hydraulic machines remains constant, that through the radial flow gas turbine changes significantly, and this change must be taken into account. One suggested volume flow rate to use is that at the outlet Q_2. Therefore referring to Eq. (1.16) and writing (gH) in terms of the enthalpy,

$$N_s = NQ_2^{0.5}/(h_{00} - h_{02ss})^{0.75} \quad (7.17)$$

Now $N = U_1/\pi D_1$ and $h_{00} - h_{02ss} = C_s^2/2$ and, upon substituting, Eq. (7.17) becomes

$$N_s = (U_1/\pi D_1 N)^{0.5}(U_1/\pi D_1)[Q_2^{0.5}/(C_s^2/2)^{0.75}]$$
$$= 0.302(Q_2/ND_1^3)^{0.5}(U_1/C_s)^{1.5} \quad (7.18)$$

Equation (7.7) showed that for an ideal turbine the ratio U_1/C_s was equal to 0.707. Therefore substituting for this into Eq. (7.18) gives

$$N_s = 0.18(Q_2/ND_1^3)^{0.5} \quad (7.19)$$

If a uniform axial velocity at exit is assumed, then $Q_2 = A_2 C_2$. Also writing the projected area of the rotor as $A_r = \pi D_1^2/4$, then

$$Q_2/ND_1^3 = \pi A_2 C_2/U_1 D_1^2 \quad (7.20)$$

Multiplying the numerator and denominator by A_r, Eq. (7.20) may be arranged as

$$Q_2/ND_1^3 = (A_2/A_r)(\pi^2/4)(C_2/U_1)$$
$$= (A_2/A_r)(\pi^2/4)(1.414/C_s)C_2$$
$$= 3.49(A_2/A_r)(C_2/C_s)$$

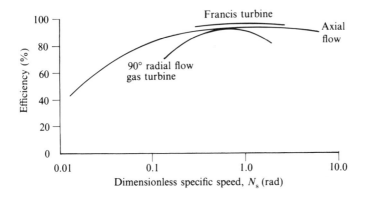

Figure 7.5 Variation of efficiency with dimensionless specific speed

Thus substituting for Q_2/ND_1^3 in Eq. (7.19) gives

$$N_s = 0.336(C_2/C_s)^{0.5}(A_2/A_r)^{0.5} \quad \text{rev} \qquad (7.21)$$

$$= 2.11(C_2/C_s)^{0.5}(A_2/A_r)^{0.5} \quad \text{rad} \qquad (7.22)$$

In practice

$$0.04 < (C_2/C_s)^2 < 0.3$$

$$0.1 < (A_2/A_r) < 0.5$$

Then

$$0.3 < N_s < 1.1 \quad \text{rad}$$

Thus the dimensionless specific speed range is very small and the variation of efficiency with N_s is shown in Fig. 7.5, where it is seen to match the axial flow gas turbine over the limited range of N_s.

EXERCISES

In the following exercises, unless otherwise stated, use the following values:
   Specific heat at constant pressure, $C_p = 1147 \, \text{J/kg K}$
   Ratio of specific heats, $\gamma = 1.333$, gas constant $R = 287 \, \text{J/kg K}$

7.1 An inward flow radial gas turbine operates with a total-to-total efficiency (from nozzle inlet to diffuser outlet) of 0.9. At entry to the nozzles, the stagnation pressure and temperature of the gas are 300 kPa and 1150 K respectively. At outlet from the diffuser the pressure is 100 kPa and the velocity of flow is negligible at that point. Find the impeller tip speed and the flow angle at the nozzle outlet if the gas enters the impeller radially and there is no whirl at the impeller exit. The Mach number at exit from the nozzles is 0.9.

7.2 A small inward flow radial exhaust gas turbine has the following design data:

Rotor inlet tip diameter	90 mm
Rotor outlet tip diameter	62 mm
Rotor outlet hub diameter	25 mm
Ratio C_2/C_s	0.447
Ratio U_1/C_s (ideal)	0.707
Blade speed	30 000 rpm
Density at impeller exit	1.8 kg/m^3

Determine:
- (a) the dimensionless specific speed of the turbine,
- (b) the volume flow rate at impeller outlet and
- (c) the power developed by the turbine.

7.3 The design data of a proposed inward radial flow exhaust gas turbine are as follows:

Stagnation pressure at inlet to nozzles, p_{00}	700 kPa
Stagnation temperature at inlet to nozzles, T_{00}	1075 K
Static pressure at exit from nozzles, p_1	510 kPa
Static temperature at exit from nozzles, T_1	995 K
Static pressure at exit from rotor, p_2	350 kPa
Static temperature at exit from rotor, T_2	918 K
Stagnation temperature at exit from rotor, T_{02}	920 K
Ratio r_{2av}/r_1	0.5
Rotational speed, N	26 000 rpm

The flow into the rotor is purely radial and at exit the flow is axial at all radii. Calculate:
- (a) the total-to-static efficiency of the turbine,
- (b) the outer diameter of the rotor,
- (c) the enthalpy loss coefficient for the nozzle and rotor rows,
- (d) the blade outlet angle at the mean diameter β_{2av} and
- (e) the total-to-total efficiency of the turbine.

7.4 Using the data of exercise 7.3, the mass flow of exhaust gas available to the turbine is 2.66 kg/s. Calculate:
- (a) the volume flow rate at rotor exit,
- (b) the hub and tip diameters of the rotor at exit,
- (c) the power developed by the turbine,
- (d) the rotor exit blade angles at the hub and tip,
- (e) the nozzle exit angle and
- (f) the ratio of rotor width at inlet to its inlet tip diameter.

7.5 An inward flow radial turbine is fitted with a diffuser, which can be assumed to have an efficiency of 100 per cent. If the turbine chokes when the absolute velocity at exit from the impeller reaches the speed of sound ($M_2 = 1$), show that the limiting total pressure ratio is given by

$$1/(1 - R^n) = (C_2/C_s)^2\{[1 + 0.5(\gamma - 1)M_2^2]/[0.5(\gamma - 1)M_2^2]\} + \eta_{t(t-t)}$$

where $R^n = (p_{02}/p_{00})^{(\gamma - 1)/\gamma}$.

SOLUTIONS

Exercise 7.1 The overall isentropic efficiency of the turbine from inlet to diffuser is given by

$$\eta_{t(t-t)} = \frac{T_{00} - T_{03}}{T_{00} - T_{03ss}}$$

Now from Eq. (7.3)

$$W/m = U_1^2 = C_p(T_{00} - T_{03})$$

when $C_{x2} = 0$. Also

$$T_{00}\left(1 - \frac{T_{03ss}}{T_{00}}\right) = T_{00}\left[1 - \left(\frac{p_{03}}{p_{00}}\right)^{(\gamma-1)/\gamma}\right]$$

Therefore

$$U_1^2 = \eta_{t(t-t)}C_pT_{00}\left[1 - \left(\frac{p_{03}}{p_{00}}\right)^{(\gamma-1)/\gamma}\right]$$

$$= 0.9 \times 1147 \times 1150\left[1 - \left(\frac{100}{300}\right)^{0.2498}\right]$$

Impeller tip speed $U_1 = 533.7\,\text{m/s}$

The Mach number of the absolute flow velocity at exit from the nozzle is given by

$$M_1 = C_1/a_1$$

$$= \frac{U_1}{a_1 \sin \alpha_1}$$

Across the nozzle the flow is adiabatic; therefore

$$T_{00} = T_{01} = T_1 + \frac{C_1^2}{2C_p}$$

$$= T_1 + \frac{U_1^2}{2C_p \sin^2 \alpha_1}$$

or

$$\frac{T_1}{T_{00}} = 1 - \frac{U_1^2}{2C_pT_{00} \sin^2 \alpha_1}$$

Now

$$C_p - C_v = R$$

and

$$1 - \frac{1}{\gamma} = \frac{R}{C_p}$$

or

$$\frac{\gamma - 1}{\gamma} = \frac{R}{C_p}$$

and substituting for C_p we get

$$\frac{T_1}{T_{00}} = 1 - \frac{U_1^2(\gamma - 1)}{2\gamma R T_{00} \sin^2 \alpha_1}$$

$$= 1 - \frac{U_1^2(\gamma - 1)}{2a_{01}^2 \sin^2 \alpha_1}$$

But

$$\left(\frac{T_1}{T_{00}}\right)^{1/2} = \frac{a_1}{a_{00}} = \frac{a_1}{a_{01}} \qquad \text{since} \qquad T_{00} = T_{01}$$

and

$$\frac{a_1}{a_{01}} = \frac{U_1}{M_1 a_{01} \sin \alpha_1}$$

Therefore

$$\left(\frac{U_1}{M_1 a_{01} \sin \alpha_1}\right)^2 = 1 - \frac{U_1^2(\gamma - 1)}{2a_{01}^2 \sin^2 \alpha_1}$$

and

$$1 = \left(\frac{U_1}{a_{01} \sin \alpha_1}\right)^2 \left(\frac{\gamma - 1}{2} + \frac{1}{M_1^2}\right)$$

$$\sin^2 \alpha_1 = \left(\frac{U_1}{a_{01}}\right)^2 \left(\frac{\gamma - 1}{2} + \frac{1}{M_1^2}\right)$$

But

$$a_{01}^2 = \gamma R T_{01}$$

$$= 1.333 \times 287 \times 1150$$

$$= 439\,956 \, \text{m}^2/\text{s}^2$$

Therefore

$$\sin^2 \alpha_1 = \frac{533.7^2}{439\,956} \left(\frac{0.333}{2} + \frac{1}{0.9^2}\right)$$

$$= 0.9071$$

Nozzle angle $\alpha_1 = 72.25°$

Exercise 7.2 (a) From Eq. (7.21) the dimensionless specific speed is

$$N_s = 0.336 \left(\frac{C_2}{C_s}\right)^{1/2} \left(\frac{A_2}{A_r}\right)^{1/2} \qquad \text{rev}$$

Now

$$A_2 = \frac{\pi(D_{2\,\text{tip}}^2 - D_{2\,\text{hub}}^2)}{4}$$

$$= \frac{\pi(0.062^2 - 0.025^2)}{4}$$

$$= 2.53 \times 10^{-3}\,\text{m}^2$$

$$A_r = \frac{\pi D_1^2}{4}$$

$$= \frac{0.09^2 \pi}{4}$$

$$= 6.361 \times 10^{-3}\,\text{m}^2$$

Dimensionless specific speed

$$N_s = 0.336 \left(\frac{0.447 \times 2.53}{6.36} \right)^{1/2}$$

$$= 0.142\,\text{rev}$$

$$= 0.89\,\text{rad}$$

If this value is compared with Fig. 7.5 it is seen that a reasonable overall efficiency is achieved.

(b) The flow rate at outlet for the ideal turbine is given by Eq. (7.19)

$$N_s = 0.18 \left(\frac{Q_2}{ND_1^3} \right)^{1/2}\,\text{rev}$$

$$0.142 = 0.18 \left(\frac{Q_2 \times 60}{30\,000 \times 0.09^3} \right)^{1/2}$$

whence

$$Q_2 = 0.227\,\text{m}^3/\text{s}$$

(c) From Eq. (7.3)

$$W = mU_1^2$$

$$= \rho_2 Q_2 U_1^2$$

$$= 1.8 \times 0.227 \times \left(\frac{\pi ND_1}{60} \right)^2$$

$$= 1.8 \times 0.227 \times \left(\frac{\pi \times 30\,000 \times 0.09}{60} \right)^2$$

Power developed $= 8.166\,\text{kW}$

Exercise 7.3 (a) The total-to-static efficiency is given by

$$\eta_{t(t-s)} = \frac{h_{00} - h_{02}}{h_{00} - h_{2ss}}$$

$$= \frac{1 - (T_{02}/T_{00})}{1 - (T_{2ss}/T_{00})}$$

From isentropic relationships

$$\frac{T_{2ss}}{T_{00}} = \left(\frac{p_2}{p_{00}}\right)^{(\gamma-1)/\gamma}$$

Therefore

$$\eta_{t(t-s)} = \frac{1 - (T_{02}/T_{00})}{1 - (p_2/p_{00})^{(\gamma-1)/\gamma}}$$

$$= \frac{1 - (920/1075)}{1 - (350/700)^{0.2498}}$$

$$= 0.144/0.159$$

$$= \underline{0.906}$$

(b) From Eq. (7.3) the specific work done is

$$W/m = U_1^2$$

since $C_{x2} = 0$ and $C_{x1} = U_1$ for radial inlet flow. Therefore

$$C_p(T_{00} - T_{02}) = U_1^2$$

and

$$U_1^2 = 1147(1075 - 920)$$

$$= 177\,785\,\text{m}^2/\text{s}^2$$

$$U_1 = 421.6\,\text{m/s}$$

Then

$$D_1 = \frac{60\,U_1}{\pi N}$$

$$= \frac{60 \times 421.6}{\pi \times 26\,000}$$

$$= \underline{0.31\,\text{m}}$$

(c) Equation (7.11) gives the nozzle loss coefficient, which we can arrange into an equation involving only pressures and temperatures:

$$\zeta_N = \frac{h_1 - h_{1s}}{0.5\,C_1^2}$$

Now

$$\frac{C_1}{2C_p} = T_{01} - T_1 = T_{00} - T_1 \qquad \text{(since } T_{01} = T_{00}\text{)}$$

Therefore

$$\zeta_N = \frac{T_1 - T_{1s}}{T_{00} - T_1}$$

We may write

$$T_{1s} = \left(\frac{T_{1s}}{T_{00}}\right) T_{00}$$

$$= \left(\frac{p_1}{p_{00}}\right)^{(\gamma - 1)/\gamma} \times T_{00}$$

$$= \left(\frac{510}{700}\right)^{0.2498} \times 1075$$

$$= 993.2\,\text{K}$$

Substituting

$$\zeta_N = \frac{995 - 993.2}{1075 - 995}$$

Nozzle loss coefficient = 0.0225

The rotor loss coefficient is given by Eq. (7.13)

$$\zeta_R = \frac{h_2 - h_{2s}}{0.5\,W_2^2}$$

Now

$$W_2^2 = C_2^2 + U_2^2$$

and using the mean diameter for the calculation of conditions at the impeller outlet

$$\frac{U_2}{U_1} = \frac{r_{2av}}{r_1}$$

$$U_2 = 0.5 \times 421.6$$

$$= 210.8\,\text{m/s}$$

Also

$$C_2^2 = 2C_p(T_{02} - T_2)$$
$$= 2 \times 1147(920 - 918)$$
$$= 4588 \, \text{m}^2/\text{s}^2$$

Therefore

$$W_2^2 = 4588 + 210.8^2$$
$$= 49\,025 \, \text{m}^2/\text{s}^2$$

and for the numerator

$$h_2 - h_{2s} = C_p \left[T_2 - \left(\frac{T_{2s}}{T_1} \right) T_1 \right]$$
$$= C_p \left[T_2 - \left(\frac{p_2}{p_1} \right)^{(\gamma - 1)/\gamma} T_1 \right]$$
$$= 1147(918 - 905.7)$$
$$= 14\,108 \, \text{J/kg (or m}^2/\text{s}^2)$$

Therefore

$$\zeta_R = \frac{14\,108}{0.5 \times 49\,025}$$
$$= 0.58$$

(d) The total-to-total efficiency is found from Eq. (7.16).

$$\frac{1}{\eta_{t(t-t)}} = \frac{1}{\eta_{t(t-s)}} - \frac{1}{2} \left(\frac{r_{2av}}{r_1} \cot \beta_{2av} \right)^2$$

Now

$$\cot \beta_{2av} = C_2/U_2$$
$$= \frac{(4588)^{1/2}}{210.8}$$
$$= 0.321$$
$$\beta_{2av} = 72.2°$$

Therefore

$$\frac{1}{\eta_{t(t-t)}} = \frac{1}{0.906} - 0.5(0.5 \times 0.321)^2$$
$$= 1.09$$
$$\eta_{t(t-t)} = 0.917$$

Exercise 7.4 The dimensionless specific speed based on the total-to-total enthalpy drop in the turbine is given by Eq. (7.17).

$$N_s = \frac{N Q_2^{1/2}}{(h_{00} - h_{02ss})^{3/4}}$$

But

$$(h_{00} - h_{02ss}) = T_{00} C_p \left(1 - \frac{T_{02ss}}{T_{00}} \right)$$

Using isentropic relations

$$(h_{00} - h_{02ss}) = C_p T_{00} \left[1 - \left(\frac{p_{02}}{p_{00}} \right)^{(\gamma - 1)/\gamma} \right]$$

Now

$$\frac{p_2}{p_{02}} = \left(\frac{T_2}{T_{02}} \right)^{\gamma/(\gamma - 1)}$$

Therefore

$$p_{02} = 350 \left(\frac{920}{918} \right)^4$$

$$= 353.06 \, \text{kPa}$$

and

$$(h_{00} - h_{02ss}) = 1147 \times 1075 \left[1 - \left(\frac{353.06}{700} \right)^{0.2498} \right]$$

$$= 193.78 \, \text{kJ/kg}$$

Rotational speed

$$\omega = \frac{2\pi N}{60}$$

$$= \frac{2\pi \times 26\,000}{60}$$

$$= 2722.7 \, \text{rad/s}$$

$$Q_2 = m/\rho_2$$

and

$$\rho_2 = \frac{p_2}{R T_2}$$

$$= \frac{350 \times 10^3}{287 \times 918}$$

$$= 1.328 \, \text{kg/m}^3$$

Therefore

$$Q_2 = \frac{2.66}{1.328}$$

$$= 2.0 \, \text{m}^3/\text{s}$$

This gives

$$N_s = \frac{2722.7 \times (2.0)^{1/2}}{(193.78 \times 10^3)^{3/4}}$$

$$= 0.416 \, \text{rad}$$

Since it was not stated in the problem which was the efficiency of importance the dimensionless specific speed could equally have been based on the total-to-static enthalpy drop $(h_{00} - h_{2ss})$ to correspond to the total-to-static efficiency. The error is small however:

$$h_{00} - h_{2ss} = C_p T_{00} \left[1 - \left(\frac{p_2}{p_{00}} \right)^{(\gamma - 1)/\gamma} \right]$$

$$= 196\,035 \, \text{J/kg}$$

and

$$N_s = 0.413 \, \text{rad}$$

(b) Volume flow rate

$$Q_2 = A_2 C_2$$

$$= \pi (r_{2\,\text{tip}}^2 - r_{2\,\text{hub}}^2) C_2$$

$$= 2\pi r_{2\text{av}} h C_2$$

where h is the height of the blade between hub and tip and $r_{2\text{av}} = (r_{2\,\text{hub}} + r_{2\,\text{tip}})/2$. Now

$$C_2^2 = 2C_p(T_{02} - T_2)$$

$$= 2 \times 1147(920 - 918)$$

$$= 4588 \, \text{m}^2/\text{s}^2$$

$$C_2 = 67.7 \, \text{m/s}$$

and

$$r_{2\text{av}} = 0.5 \times 0.155$$

$$= 0.0775 \, \text{m}$$

Therefore

$$h = \frac{2.0}{2\pi \times 0.0775 \times 67.7}$$

$$= 0.061 \text{ m}$$

$$r_{2\,\text{hub}} = r_{2\text{av}} - \frac{h}{2}$$

$$= (0.5 \times 0.155) - \frac{0.061}{2}$$

$$= 0.047 \text{ m}$$

Hub diameter $= 0.094$ m

$$r_{2\,\text{tip}} = r_{2\text{av}} + \frac{h}{2}$$

$$= (0.5 \times 0.155) + \frac{0.061}{2}$$

$$= 0.108 \text{ m}$$

Tip diameter $= 0.216$ m

(c) From Eq. (7.3)

$$W = mU_1^2$$

$$= 2.66 \times 421.6^2$$

Power developed $= 472.8$ kW

(d) From the outlet velocity triangle, since C_2 is axial and uniform over the exit annulus,

$$\tan \beta_2 = U_2/C_2$$

At the tip

$$U_{2\,\text{tip}} = U_1 \frac{r_{2\,\text{tip}}}{r_1}$$

$$= 421.6\left(\frac{0.108}{0.155}\right)$$

$$= 293.8 \text{ m/s}$$

$$\tan \beta_2 = \frac{293.8}{67.7}$$

At exit tip $\beta_2 = 77°$

At the hub

$$U_{2\,\text{hub}} = 421.6\left(\frac{0.047}{0.155}\right)$$

$$= 127.8\,\text{m/s}$$

$$\tan\beta_2 = \frac{127.8}{67.7}$$

At exit hub $\beta_2 = 62°$

(e) Across the nozzle $h_{00} = h_{01}$ and

$$C_1^2 = 2C_p(T_{00} - T_1)$$

$$= 2 \times 1147(1075 - 995)$$

$$= 183\,520\,\text{kJ/kg}$$

Nozzle exit velocity

$$C_1 = 428.4\,\text{m/s}$$

The inlet velocity triangle shows

$$\sin\alpha_1 = U_1/C_1$$

$$= \frac{421.6}{428.4}$$

Nozzle angle $\alpha_1 = 79.7°$

(f) From the continuity equation at impeller inlet,

$$m = \rho_1 A_1 C_{r1}$$

$$= \rho_1 \pi D_1 W_1 b_1$$

where b_1 is the depth of the impeller blades at inlet

$$\frac{b_1}{D_1} = \frac{m}{\pi \rho_1 D_1^2 W_1}$$

But

$$W_1 = U_1 \cot\alpha_1$$

$$= 421.6 \cot 79.7°$$

$$= 76.62\,\text{m/s}$$

and

$$\rho_1 = \frac{p_1}{RT_1}$$

$$= \frac{510}{287 \times 995}$$

$$= 1.786\,\text{kg/m}^3$$

Therefore substituting

$$\frac{b_1}{D_1} = \frac{2.66}{\pi \times 1.79 \times 0.31^2 \times 76.62}$$

$$= 0.0642$$

Blade depth $= 0.02\,\text{m}$

Exercise 7.5 Since the flow is isentropic in the diffuser, $T_{02ss} = T_{03ss}$, $T_{02} = T_{03}$ and $p_{02} = p_{03}$. With the diffuser the total-to-total efficiency is

$$\eta_{t(t-t)} = \frac{T_{00} - T_{02}}{T_{00} - T_{02ss}}$$

From Table 7.1 the spouting velocity is

$$\frac{C_s^2}{2C_p} = T_{00} - T_{02ss}$$

Substituting

$$\eta_{t(t-t)} = \frac{1}{1 - (T_{02ss}/T_{00})} - \frac{2T_{02}C_p}{C_s^2} \tag{i}$$

Using isentropic relationships

$$\frac{T_{02ss}}{T_{00}} = \left(\frac{p_{02}}{p_{00}}\right)^{(\gamma - 1)/\gamma}$$

Substituting for the temperature ratio and multiplying both sides by C_2^2 we get

$$C_2^2 \eta_{t(t-t)} = \frac{C_2^2}{1 - (p_{02}/p_{00})^{(\gamma - 1)/\gamma}} - 2T_{02}C_p \left(\frac{C_2}{C_s}\right)^2$$

or

$$\left(\frac{C_2}{C_s}\right)^2 = \left(\frac{1}{1 - (p_{02}/p_{00})^{(\gamma - 1)/\gamma}} - \eta_{t(t-t)}\right)\frac{C_2^2}{2C_p T_{02}} \tag{ii}$$

Now

$$\frac{T_{02}}{T_2} = 1 + \frac{C_2^2}{2C_p T_2} \qquad \text{(iii)}$$

and

$$\frac{R}{C_p} = \frac{C_p - C_v}{C_p} = 1 - \frac{1}{\gamma} = \frac{\gamma - 1}{\gamma}$$

Substituting for C_p in Eq. (iii)

$$\frac{T_{02}}{T_2} = 1 + \frac{(\gamma - 1)}{2} \frac{C_2^2}{\gamma R T_2}$$

But

$$a_2^2 = \gamma R T_2$$

Therefore

$$\frac{T_{02}}{T_2} = 1 + \left(\frac{\gamma - 1}{2}\right) M_2^2 \qquad \text{(iv)}$$

Thus

$$\frac{C_2^2}{2C_p} = T_{02} - T_2$$

$$= \frac{T_2(\gamma - 1)}{2} M_2^2 \qquad \text{(v)}$$

and hence

$$\frac{C_2^2}{2C_p T_{02}} = \frac{\frac{1}{2}(\gamma - 1)M_2^2}{1 + \frac{1}{2}(\gamma - 1)M_2^2} \qquad \text{(vi)}$$

Substituting from Eq. (vi) into Eq. (ii) and writing $R^n = (p_{02}/p_{00})^{(\gamma-1)/\gamma}$

$$\left(\frac{C_2}{C_s}\right)^2 = \left(\frac{1}{1 - R^n} - \eta_{t(t-t)}\right)\left(\frac{\frac{1}{2}(\gamma - 1)M_2^2}{1 + \frac{1}{2}(\gamma - 1)M_2^2}\right)$$

and rearranging

$$\frac{1}{1 - R^n} = \left(\frac{C_2}{C_s}\right)^2 \left(\frac{1 + \frac{1}{2}(\gamma - 1)M_2^2}{\frac{1}{2}(\gamma - 1)M_2^2}\right) + \eta_{t(t-t)}$$

or

$$\frac{1}{1 - (p_{02}/p_{00})^{(\gamma-1)/\gamma}} = \left(\frac{C_2}{C_s}\right)^2 \left(\frac{1 + \frac{1}{2}(\gamma - 1)M_2^2}{\frac{1}{2}(\gamma - 1)M_2^2}\right) + \eta_{t(t-t)}$$

For chosen values of $\eta_{t(t-t)}$ and M_2, $(C_2/C_s)^2$ can be varied and (p_{02}/p_{00}) determined.

REFERENCES

1. Stodola, A., *Steam and Gas Turbines*, vols I and II, McGraw-Hill, New York, 1927. (Reprinted, Peter Smith, New York, 1945.)
2. Wislicenus, G.F., *Fluid Mechanics of Turbomachinery*, McGraw-Hill, New York, 1947.
3. Stanitz, J.D., 'Some theoretical aerodynamic investigations of impellers in radial and mixed flow centrifugal compressors', *Trans. ASME*, vol. 74, p. 4, 1952.
4. Stepanoff, A.J., *Centrifugal and Axial Flow Pumps*, Wiley, New York, 1948.
5. Jacobs, E.N., Ward, E. and Pinkerton, R.M., 'The characteristics of 78 related aerofoil sections from tests in the variable density wind tunnel', NACA Report No. 460.
6. Soderberg, C. R., Unpublished note, Gas Turbine Laboratory, Massachusetts Institute of Technology, 1949.
7. Ainley, D.G. and Mathieson, G.C.R., 'A method of performance estimation for axial flow turbines', A.R.C., R and M 2974, 1951.
8. Benson, R.S., 'A review of methods for assessing loss coefficients in radial gas turbines', *Int. J. Mech. Sci.*, vol. 12, 1970.

INDEX